FIELD EFFECT
IN SEMICONDUCTOR-ELECTROLYTE
INTERFACES

FIELD EFFECT
IN SEMICONDUCTOR-ELECTROLYTE
INTERFACES

*Application to Investigations of
Electronic Properties of
Semiconductor Surfaces*

PAVEL P. KONOROV,
ADIL M. YAFYASOV,
AND VLADISLAV B. BOGEVOLNOV

PRINCETON UNIVERSITY PRESS
PRINCETON AND OXFORD

Library of Congress Cataloging-in-Publication Data
Konorov, P. P.
Field effect in semiconductor-electrolyte interfaces: application to
investigations of electronic properties of semiconductor surfaces /
Pavel P. Konorov, Adil M. Yafyasov, and Vladislav B. Bogevolnov.
p. cm.
Includes bibliographical references and index.
ISBN-13: 978-0-691-12176-5 (cloth: alk. paper)
ISBN-10: 0-691-12176-1 (cloth: alk. paper)
1. Semiconductors—Junctions. 2. Electrochemical analysis. I. Yafyasov,
Adil M., 1951– II. Bogevolnov, Vladislav B., 1954– III. Title.
QC611.6.J85K66 2006
537.6'22—dc22   2005055241

British Library Cataloging-in-Publication Data is available

This book has been composed in Utopia Typeface

Printed on acid-free paper. ∞

pup.princeton.edu

Printed in the United States of America

1 3 5 7 9 10 8 6 4 2

◇ Contents ◇

⬦ *Preface* ⬦

T̲HE SUBJECT MATTER of this book is the interface between two phases: a semiconductor and an electrolyte. Interfaces are of particular interest from the point of view of understanding a universal subsystem of our world, since they divide materials with different electronic properties. These interface domains are localized within the near-surface region of separated charges and may be considered special states of the condensed phase which in principle are unattainable within the bulk phase. Interfaces, including surfaces, are highly active domains, decisive for the character and mode of behavior of most processes taking place in nature. Accordingly, consideration of questions connected with the role of interfaces comes within the scope of different subjects in the natural sciences such as physics, chemistry, biology, medical science, and so on.

In physics approximations the states of interfaces are represented in the form of boundary data, which fix the basic regularities of the electrophysical properties of the solid.

In electronics, almost all the electronic processes determining the functioning of solid electronic devices are those at interfaces (p-n junctions, heterostructures, semiconductor-insulator junctions, metal-insulator-semiconductor (MIS) structures, Schottky diodes, and so on). In particular, processes at interfaces concern objects of nanometer size, which are at this time of the greatest interest for both physics and technology.

In chemistry interfaces are the most active domain where different chemical reactions take place. The fundamental role of electrical properties of interfaces is apparent in their unique catalytic activity.

In living organisms different interfaces, including cellular membranes, are regions through which the exchange of energy and matter with the surroundings takes place as a necessary condition for their existence. Naturally these processes are of great interest for biology and medical science.

Interfaces are inherent in open systems and the only barriers on the path of energy flows and particle streams that appear under the influence of density gradients and superimposed fields. It is reasonable that preponderance of the field-controlled barriers over the unalterable ones is a governing factor among the many conditions that are necessary to realize self-organizing phenomena. These phenomena play an important role in complicated systems of all kinds from the physical to the biological and are fascinating to scientists, because new knowledge about elementary processes in these systems allows other than thermodynamic descriptions to be developed.

In particular, it is necessary to point out that interfaces separate solid and liquid phases whose properties are determined by interaction of the electronic subsystems in solid and ionic forms of electrolytes. The properties of the interfaces in fact determine the character and nature of the electrochemical reactions and are the basis of electrochemistry. They play an important role in processes that take place in biological systems including different living organisms because the liquid in cellular space contains ions and is analogous to an electrolyte.

So one of the important problems of natural science is to find some common rules that can be generalized from the interface properties in different systems and to settle general laws ruling the processes in such systems. Solving this problem is important in both theory and practice and is possible only by studying results obtained in different spheres of knowledge.

The basic functions of interfaces and the nature of processes in them are determined by their charge state, that is, by the value and space distribution of separated charges in the interface region, and also by its change under the influence of different external factors that result in the redistribution of charge in the region of the interface. Charge distribution is a universal property of all interfaces and for its study it is necessary to use methods sensitive to the transfer and accumulation of charge in the interface state.

In the case of semiconductors one such method is the field effect, in which the charge in the semiconductor-insulator interface is changed by application of an external electrical field to a MIS structure. This effect allows control of the charge processes in a semiconductor-electrolyte interface and is also the base for important practical applications of MIS structures in electronics and as a basic method of semiconductor surface parameter investigation.

This monograph is devoted to consideration of the variety of the field-effect method realized in semiconductor-electrolyte (SE) systems by application of an external electric field, and, to distinguish it from the MIS technique, it is hereafter called the field effect in the SE (FESE) system. The peculiarities of the SE system allow one to make good use of this effect, on the one hand, as a way to change or control the charge state of semiconductor surfaces and interfaces, and, on the other hand, as a new method for investigation and determination of their electronic characteristics (FESE method).

At the same time the processes of charge state change that take place under electric field action are universal and may occur in various phenomena, including light effects, and atomic or molecular adsorption from an electrolyte, for both chemical and electrochemical processes, and other events within interfaces. That is why rules obtained with the help of the FESE method may apply to a wide range of processes in different systems, varying from physics to biology and medical science. A general study of these charge processes may provide a clue for understanding phenomena exhibiting complicated behavior. Therefore the processes and relationships displayed by the field effect in semiconductor-electrolyte interfaces may be interesting both for specialists in semiconductor surface and interface physics as well as semiconductor electrochemistry and for researchers who are engaged in questions connected with electronic-ionic processes.

This book is specialized in nature and intended for those who are specializing in surface physics, the science of materials, and electronic engineering. Nevertheless, the authors have made an attempt to make the book intelligible for those who are interested in the field. It may be used in universities and research centers and may be useful for teachers, students, and engineers working in corresponding spheres.

ACKNOWLEDGMENTS

The authors would like to thank Dr. A. Cherednichenko for his helpful comments in preparing the manuscript and I. Ivankiv for technical assistance. This book is dedicated to the blessed memory of Ilya Romanovich Prigogine.

SCL   space charge layer
SE   semiconductor-electrolyte
MIS   metal-insulator-semiconductor
FE   field effect
FESE   field effect in a semiconductor-electrolyte interface
QW   quantum wire
HTSC   high-temperature superconductor
MCT   mercury cadmium telluride
EDTA   ethylene diamine tetra-acetate

Introduction

THE SPECIAL INTEREST in semiconductor surface electronic and atomic processes for the study and understanding of associated mechanisms is largely due to the complicated and universal nature of the surface structure. A real semiconductor surface is actually a multilayer system. It consists of an outer, "foreign" layer which is usually represented by the oxide of the material, and also of a semiconductor layer which, in its turn, can be divided into two neighboring layers. The first one has a disturbed crystalline structure; the other features bulk crystalline properties but its electric state differs; this layer is referred to as the space charge layer (SCL). Electronic and atomic processes occurring in these layers are interdependent; the processes in each of these layers as well as the interrelation between them are of primary importance for surface and near-surface phenomena. As a result, a real semiconductor surface is a complex system which should be treated as an open thermodynamic one where processes have, for the most part, a nonequilibrium and irreversible character and require a dynamical description. The investigation of such processes requires further development of adequate approaches and techniques allowing variation of external conditions and in situ measurements under controlled changes of surface parameters and should be extended to include various different kinds of systems and interfaces.

A semiconductor-electrolyte system, hereafter referred to as a SE system is an example of a structure that satisfies these requirements. By making use of such a system, it is possible to control the change of a surface and the processes occurring on it. This can be achieved by means of controlled variations of the electrolyte composition as well as chemical and electrochemical reactions, the value and sign of the polarization field, adsorption of various surface-active particles, application of electric and magnetic fields, and light illumination through a transparent electrolyte contact [1–5]. The utilization of this system provides a further extension of the techniques for surface studies permitting in situ measurements, which include, apart from traditional methods, those typical for electrochemical studies on electrode surfaces [6, 7].

One of the advantages of the SE system is the possibility of using one of the modifications of the field effect (FE). This is known to be a useful technique for study of semiconductor surfaces with the help of metal-insulator-semiconductor (MIS) structures [8]. In a general sense, the FE technique can be regarded as a means for alteration of surface electronic properties as well as those of near-surface regions due to charge redistribution at the "semiconductor" side of the MIS capacitor under the application of an external electric field. In this formulation, the FE technique includes measurements of such parameters as the surface capacitance and conductance, the surface recombination velocity, the photopotential, and the carrier mobility. Measurements of these parameters and of their dependencies on the value of the external electric field yield such characteristics as the surface potential V_s (band bending) and the density and energy distribution of electronic states on the surface and in the near-surface region of the semiconductor. The type of free-carrier scattering in the near-surface region as determined by the band bending and surface states can also be found.

The feasibility of FE realization in a SE system, referred to as the field effect in the semiconductor-electrolyte (FESE) interface, is due to the fact that the SE system can be regarded as a field capacitor in which the electrolyte plays the role of a metal electrode whereas the insulator interlayer is the so-called Helmholtz layer; its thickness is 3–5 Å and is determined by the distance from the semiconductor surface to the center of the electrolyte ion electrostatically adsorbed on the semiconductor [1, 3, 9]. The semiconductor polarization in the electrolyte relative to an additional electrode results in a change of the value and sign of the ionic charge on the surface, thus enabling controlled alteration of the surface band bending. These changes correspond to changes of the electrode potential of the semiconductor electrode, which is measured relative to the reference electrode immediately in the electrolyte. Due to the small thickness and high degree of homogeneity of the Helmholtz layer and, as a consequence, the high capacitance of the field capacitor, values of the areal charge density of the order of 10^{-5} C $\cdot$ cm^{-2} may be attained even on application of a voltage as low as a few volts. This enables the development of high electric field strengths near the surface which, accordingly, produce band bending normally unattainable in conventional MIS structures. In this case, degeneracy and low-dimensional effects in the electron gas in narrow-gap and gapless semiconductors arise. This allows the basic ideas of phemenological theory of the SCL to be extended into a new class of materials, namely, that of gapless semiconductors and semi-metals; and the FESE can be used for obtaining novel information on the fundamental parameters of the band structure of semiconductor near-surface regions.

The reviews and monographs on semiconductor electrodes so far published were mainly concerned with electrochemical aspects of a semiconductor-electrolyte interface [1–4, 10–32]. The practical significance of those studies was mainly related to such problems as cleaning of semiconductor surfaces and production on them of passivating coatings. With further advance in the field, much attention was paid to photoelectrochemical phenomena, due to the potentialities of semiconductor employment for solar energy consumption and conversion and obtaining of hydrogen fuel [17, 21–25, 30–37]. The aforementioned works also discussed the problems of the physics of electronic processes occurring at the SE boundary, since the elementary acts involved in these processes are of a physical nature. The main issues important for progress in understanding the mechanisms underlying the electrochemical processes have been discussed in pioneering works [2, 4, 5, 25, 31, 33, 35, 38–44]. These studies demonstrate the exictence of the field effect at the SE interface; they also demonstrate the potentialities of the FE technique as applied to studies of semiconductor surface properties. Yet there is no systematic description of the physical nature of the processes on semiconductor electrode surfaces based upon the unique notions of modern semiconductor surface physics. The present book is intended to partially fill this gap.

In Chapters 1 and 2 basic knowledge about the SE interface is presented. Relevant notions and terms are introduced. Principles of processes occurring under polarization which govern current-voltage $[I(\varphi)]$ characteristics, that is, polarization dependences as well as several features of FESE are treated. Conditions for the ideal polarizability of semiconductor electrodes which determine the possibilities of FESE realization in equilibrium (quasiequilibrium FESE) are considered.

In Chapter 3 we discuss basic ideas concerned with the quasiequilibrium FESE which is analogous to the conventional FE in MIS structures. The potentialities of this effect for determination of the semiconductor surface potential and other SCL characteristics are demonstrated, as exemplified by Ge. There are discussed the density of surface states and their energy spectrum which can be determined from measurements of the surface capacitance and conductance versus the electrode potential in the region of ideal polarizability.

Nonequilibrium depletion is shown to be one of the manifestations of deviation from the conditions for quasi-equilibrium FESE in Chapter 4. It is shown that in the SE system a steady nonequilibrium depletion state can be realized not only in wide-band-gap semiconductors, which are characterized in equilibrium by a rather low density of minority carriers in the bulk, but also in fairly narrow-gap semiconductors. In this case, a rather thin sample is similar to a charged insulator where high electric field strengths are realized. A system in this state offers new possibilities for investigation of various phenomena associated with the effects of strong fields on semiconductors. In this state FESE characteristics are sensitive to any of the processes of minority-carrier generation in semiconductors. This allows use of the depletion state for studies of generation and recombination processes as well as for different practical applications.

Chapter 5 is concerned with the use of the FESE technique for control of binary and multicomponent semiconductor surfaces and investigation of their electrophysical characteristics based on the possibility of extension of the semiconductor ideal polarizability region. An important consequence of the extension of the polarization region corresponding to the ideal polarizability is the possibility of obtaining, for narrow-band and gapless semiconductors, band bendings when the Fermi level at the surface lies deep (about 100–300 meV) in the allowed semiconductor bands. Consideration of the effects of degeneracy of the electron gas system occurring at the semiconductor surface shows that the utilization of the quasi-equilibrium FESE in such conditions allows the determination of the near-surface region characteristics, including densities of electron and hole states, the dispersion law, and the electron and hole effective masses. For narrow-gap semiconductors the possibility of the observation of phenomena arising due to size (two-dimensional) effects in the electron gas is shown. These effects manifest themselves in the FESE as a number of steps in the curve representing the dependence of the capacitance and conductance on the electrode potential. The possibility of observation of this phenomenon at room temperature is a consequence of the high field strength at the semiconductor surface in the SE system. In this region of electrical field the energy separation between the different quantum subbands may well exceed the value by which the quantum subbands are smeared and broadened due to temperature and collision effects. These phenomena are considered in more detail in Chapter 7.

In Chapter 5 we show also that the FESE technique advanced and developed for investigation of narrow-gap and gapless semiconductors, which makes use of the concept of the electronic state density near the Fermi level, may prove to be effective for studies of electronic properties of surfaces of metal electrodes and high-temperature superconducting (HTSC) materials. This is exemplified by investigations performed on mercury electrodes. It is shown that, under certain conditions (the choice of the electrolyte, the state of the polarization), a forbidden energy band for electrons in the near-surface mercury layer appears. This corresponds to the metal-semiconductor transition, which can be accounted for by the atomic reconstruction of the semiconductor surface layer at the semiconductor-electrolyte boundary. One of the most interesting results obtained in investigations of the FESE on the surface of HTSC materials was the observation of a negative capacitance. This effect may be treated as evidence of the inductance arising due to the piezoelectric effect on the HTSC sample surfaces under their polarization in electrolytes.

The next part of the book (Chapter 6) deals with further development of the FESE technique with an account of some specific features of the charge dynamics in the electric double layer at the SE interface. These are associated with atomic (ionic) processes which are stimulated by polarization and/or adsorption of particles from the electrolyte. Of most interest here is the effect of formation of a number of quasistable states of the semiconductor-electrolyte interface. The existence of several quasistable (bistable) states reflecting the self-consistent interaction of the two subsystems (electrons in the semiconductor and ions in the electrolyte) suggests that, as the energy is dissipated, self-organized phenomena may arise similar to those observed earlier in other systems [45]. Effects that can be interpreted as self-organized phenomena are observable at the SE interface under the combined action of polarization and adsorption of organic and polymer molecules.

The size quantization of the electron gas in the semiconductor-electrolyte interface and its manifestation in the field effect are considered in Chapter 7.

Chapter 8 is devoted to the possibilities of the FESE method for technological applications. In this chapter we also discuss the limits of employment of the quasi-equilibrium FESE technique and some questions connected with its practical realization.

In the conclusion the potentialities of the FESE and its specific features are summarized. We give perspectives and possibilities of other electrophysical techniques that could also be used for studies of semiconductor surface and interphase boundaries in a SE system and could be of interest, as well as supplementary to the FESE.

FIELD EFFECT
IN SEMICONDUCTOR-ELECTROLYTE
INTERFACES

Semiconductor-Electrolyte Interface:
Basic Notions and Definitions

A CHARACTERISTIC FEATURE of a SE system in equilibrium is the presence of exchange currents of electrochemical origin. These arise due to charge transfer across the electrochemical barrier that exists at the SE interface. The charge exchange arises as a result of chemical dissolution of the semiconductor (ionic exchange), as well as that between the electrons of the semiconductor and the ions of the electrolyte [30–32, 46–48].

It is well known that covalent-type semiconductors like Ge and Si do not dissolve in oxidizer-free aqueous electrolytes [49]. Charge exchange in this case occurs due to the exchange between the electrons of the c and v bands of the semiconductor and the ions of the electrolyte. In accordance with the Franck-Condon principle, this exchange is accomplished by electron tunneling and becomes possible in the case when the electron energy level in the "ionic" state in the electrolyte coincides with that in the semiconductor. In this case, it is assumed that the conditions allowing the electron exchange are attained as a result of thermal fluctuations in the polar electrolyte which change the energy of electrons in the "ionic" states. It is also assumed that the probability of electron transfer as such is about unity (when the Franck-Condon principle, in the conditions typical of oxidoreduction processes, holds true [21, 22, 46]). So the electron transfer is regarded as a process that is realized by means of electron exchange between states with coinciding energy levels. Thus the exchange rate between occupied and unoccupied electronic states is a function of their relative energy-level positions.

In the general case, one should expect that both electron as well as hole currents will cross the interface. The current values are proportional to the concentrations of ions in the electrolyte capable of exchanging with the electrons of the semiconductor, to the concentration of free carriers in the semiconductor at the SE interface, and to the probability of their transition across the barrier. The relevant contributions of c and v bands depend on which of those dominates the electron exchange. The relation between the corresponding currents in equilibrium is given by the following expression:

$$j_n^0/j_p^0 \approx \exp(qV_s^0/k_0 T), \qquad (1.1)$$

where V_s^0 is the equilibrium value of the potential drop in the semi-conductor SCL which defines the near-surface band bending and is a function of the oxidoreduction potential of the system. In strong oxidizer solutions the energy bands are bent upward ($V_s^0 < 0$); hence, the v band plays the major role. In the opposite case ($V_s^0 > 0$), the exchange is primarily with the c band [3, 9, 11, 21]. The occurrence of intense exchange currents governing thermodynamic equilibrium accounts for the high degree of stability of semiconductor surfaces in an electrolyte.

As a result of charge exchange a double layer will be built up. On the semiconductor side this is due to the charge in the SCL and surface states, whereas on the electrolyte side an equal charge of opposite sign will be set. The latter is due to electrolyte ions at the semiconductor surface as well as those in a thin diffuse near-surface layer in the immediate vicinity of the surface (the Goui layer) [50–52]. The potential drop in the electric double layer, referred to as the Galvani potential [9, 52], occurs mainly within the SCL of the semiconductor, whereas in the electrolyte (in the Goui layer) it can be neglected.[1]

The value of the Galvani potential can be represented as a sum of different potentials corresponding to the Fermi level position in a semiconductor, the oxidoreduction potential of the electrode reaction, and an arbitrary additive constant [9, 16]. This value is incorporated into the electrode potential (φ) which represents the potential difference between the two identical metal ends of the electrochemical circuit, the semiconductor being one of its components.[2] The electrochemical circuit also includes the reference electrode, the potential at which is considered to be zero (see Fig. 1.1). It is commonly supposed that a positive shift of φ corresponds to the transport of positive charges from the semiconductor to the electrolyte [17]. Hence, the more positive the charge and the more

---

[1] The ion concentrations in the electrolyte are assumed to be large ($N > 0, 1$ mol), with the density of surface states on the semiconductor not so great.

[2] If there are no special reservations all values of the electrode potential presented within this book were measured relative to the normal hydrogen electrode.

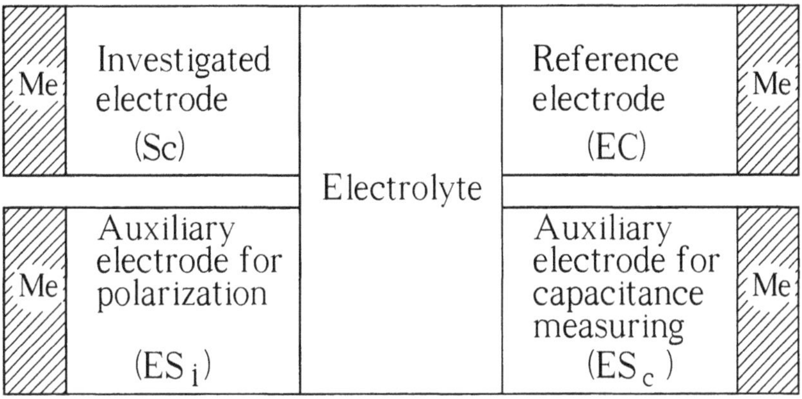

Me	Investigated electrode (Sc)		Reference electrode (EC)	Me
Me	Auxiliary electrode for polarization (ES$_i$)	Electrolyte	Auxiliary electrode for capacitance measuring (ES$_c$)	Me

Figure 1.1 Scheme of the normally broken electrochemical circuit.

negative the electrostatic potential in the SCL of a semiconductor, the more positive is its electrode potential. The value of the equilibrium electrode potential (φ_0) is a function of the chemical potentials of the electrolyte ions. Unlike the Galvani potential, the equilibrium electrode potential does not depend upon the Fermi level position in the semiconductor, because the potential drop at the Ohmic contact required for electrode potential measurements has the opposite dependence on the Fermi level. Thus, the electrode potential for a given semiconductor and electrolyte is constant independent of the incorporated impurities provided the chemical character of the semiconductor and its surface layer are not changed. Hence, measurements of the electrode potential prove to be a convenient method of monitoring the surface conditions of a semiconductor, enabling control of the degree to which the surface is "clean" and reproducible.

One of the most important characteristics useful for the description of electrochemical and physicochemical processes at the SE interface is the energy structure of the latter. To plot the interface energy scheme, assume that the electrolyte contains so-called oxidoreduction pairs, hereafter referred to as redox pairs. These can be represented, in the simplest case, by singly and doubly charged metal ions (M^+ and M^{++}) which can be regarded as occupied and unoccupied electronic states, respectively. In the first approximation, approaches similar to those employed for the treatment of donors and acceptors in semiconductors can be applied in this case as well. The electron energy levels of ions in the electrolyte may be treated as

3

the surface state levels in the conventional electron scheme of a semi-conductor. Henceforth, their energy positions are referred to as E_{ox}^0 and E_{red}^0 for occupied and unoccupied electronic states, respectively. Note the following peculiarities of these states in which they differ from the usually treated surface electronic states [8, 10, 22, 52] :

1. Ions electrostatically adsorbed on a semiconductor surface have a solvation shell which defines the universal character of the electric double layer. This simplifies consideration of electron exchange processes and enables the controlled alteration of surface band bending under realization of FESE (see Chapter 3).

2. For one and the same adsorbed ion, different charge states correspond to different energy levels of the electron states. This is stipulated by the polarization of the medium and reorientation, with ion charge state modification, of the electrolyte dipole molecules surrounding the ion. This difference is described by the reorganization (rearrangement) energy E_R, which is the energy change of a polar medium due to changes in the polarization accompanying the electron exchange between the semiconductor and electrolyte and satisfies the condition $E_{ox} - E_{red} = 2E_R$.

3. For the electron energy states of ions adsorbed on the semiconductor surface, one should take into account their energy smearing due to thermal fluctuations of ions in the electrolyte. The distribution of the probability of the level to have the energy E is given by the following expression:

$$W(E) = \sqrt{4\pi E_R k_0 T} \cdot \exp\left(-\frac{(E_t - E)^2}{4 E_R k_0 T}\right) \qquad (1.2)$$

where E_t is the energy corresponding to the distribution maximum at which the free energy of the whole system is minimized [10]. In this case the level energy exhibits fluctuations around the most probable value, which allows use of notions such as the fluctuation density of electron states in the electrolyte. Note the difference between the concept of fluctuation in the electron distribution due to the electron-phonon coupling at a given energy-level distribution and that of fluctuation of energy levels themselves, relating to electronic states in the electrolyte, which results from the ion-phonon interaction.

In considering the SE interface energy structure in equilibrium, it is appropriate to introduce the redox potential V_{redox}, which is defined as

$$- qV_{redox} = (E_{ox} + E_{red})/2 + k_0 T \ln(C_{ox}/C_{red}) \qquad (1.3)$$

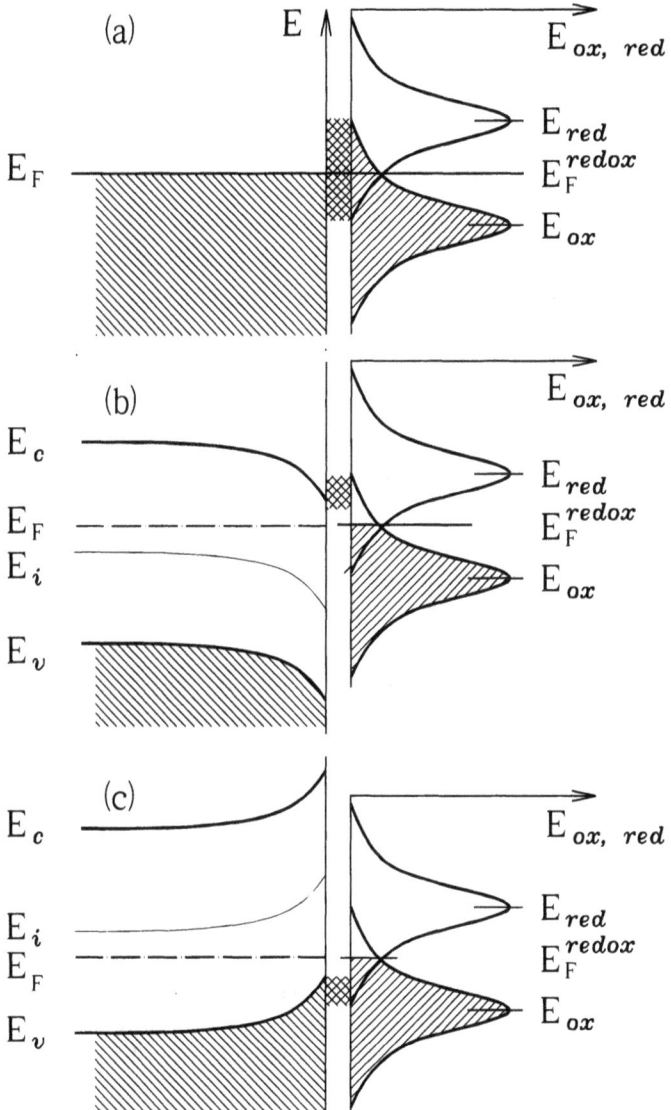

Figure 1.2 Schematic diagram of electronic energy levels of solid–electrolyte interface in thermodynamical equilibrium state: (a) metal; (b) semiconductor of n type (electron exchange with c band); (c) semiconductor of p type (electron exchange with v band). The energy interval $\sim k_0 T$ the electron exchange between solid and electrolyte occurs is double shaded.

where C_{ox} and C_{red} are the densities (activities in the more general case) of the relevant ions. Here $\ln(C_{ox}/C_{red})$ defines a correction for their concentration (activity) difference. This potential may be treated as the degree to which the redox pairs are capable of giving up or accepting electrons. With $C_{ox} = C_{red}$, the equilibrium corresponds to a coincidence of the $-qV_{redox}$ level with the Fermi level in the semiconductor, that is, $-qV_{redox} = E_F$. So the value of $-qV_{redox}$ itself may be regarded, in relation to the electron exchange, as an analogue to the Fermi level in the electrolyte. It is defined as E_F^{redox}.[3]

Under such assumptions, the energy diagram of the SE boundary can be represented as in Fig. 1.2. From the figure it follows that the major parameter defining the electron state energy distribution at the boundary between a semiconductor and an electrolyte containing a redox pair is the Fermi level position in the redox system (E_F^{redox}) with respect to the semiconductor band edges. If E_F^{redox} lies closer to the c band edge, we have the case when the exchange with the electrons of the c band dominates. When E_F^{redox} lies closer to the v band edge, the electron exchange with the v band plays the dominant role. The positions of the edges of the allowed bands at the surface of a semiconductor that is in contact with the electrolyte can be determined by finding the values of the electrode potential in relation to the reference electrode in the condition of flat bands (φ_{fb}).

Another alternative is measurement of the differential capacitance of the SE interface at electrode potentials corresponding to the degeneracy of the electron (hole) gas in the semiconductor SCL (see chapter 5).

---

[3] This analogy proceeds only from the thermodynamical relations defining equilibrium conditions at the SE interface. Strictly speaking, the position of E_F^{redox} in the electrolyte is not connected with free electrons, but is rather defined by the oxidoreduction system, containing only bound electrons [5, 10, 53].

Semiconductor-Electrolyte Interface under Polarization: Voltage-Current Relationships (Polarization Characteristics)

APPLICATION OF AN EXTERNAL VOLTAGE to a semiconductor-electrolyte phase boundary (the polarization of the semiconductor with reference to the auxiliary electrode) yields a modification of the density and character of the charge carriers involved in the transport across the interface. Correspondingly, certain changes will occur in the double electric layer at the boundary. On the electrolyte side, these result from modification of the character and alteration of the density of ions electrostatically adsorbed at the semiconductor surface. On the semiconductor side, the relevant changes are associated with changes of surface free-carrier (electrons and holes) densities arising from changes of the surface potential (band bending). Simultaneously, the relative positions of electron state energy levels are changed both in the semiconductor and in the electrolyte. This is the result of the voltage potential drop in the Helmholtz layer, which yields a modification of the intervals corresponding to the range within which the electron exchange between a semiconductor and electrolyte is available (see Chapter 1). As a consequence, the equilibrium electron exchange is disturbed, which yields the resulting current. As this takes place, the electrode potential value is different in that it becomes distinct from the equilibrium value; and this difference is equal to the alteration of the potential drop at the interface. In the case of a semiconductor, the difference is mainly associated with the potential change within the SCL, that is, we have the relation[1] $\Delta \varphi = (\varphi - \varphi_0) = -\Delta V_s$. The value of the charge density as a function of the

---

[1] The different signs of $\Delta \varphi$ and ΔV_s are due to the different directions along which the potentials are calculated according to standard practice in electrochemistry and in semiconductor surface physics.

applied voltage will dictate the general shape of the voltage-current curve of the SE boundary. Yet, in electrochemistry, it is customary to examine the dependence $\varphi(j)$, which is referred to as the polarization characteristic of the SE interface.

For metals, the disturbance of the equilibrium charge exchange at the phase boundary yielding the resulting current is due to potential drop changes in the Helmholtz layer, which plays the role of the potential barrier in the electron exchange. For a semiconductor, the appearance of the resulting current is mainly due to surface potential alterations, resulting in changes of the densities of holes and electrons involved in electron exchange. Under polarization, in the absence of degeneracy, these densities can be given in the form

$$n_s = n_s^0 \exp(q\Delta V_s/k_0 T), \tag{2.1}$$

$$p_s = p_s^0 \exp(-q\Delta V_s/k_0 T) \tag{2.2}$$

where n_s^0 and p_s^0 are the densities of electrons and holes at the semiconductor surface in equilibrium and $\Delta V_s = V_s - V_s^0$ is the surface potential change under polarization.[2] The relevant expressions for electron and hole currents across the SE interface are given in the form [38]

$$j_n^c = j_0^c \cdot [\exp(q\Delta V_s/k_0 T) - 1] = j_0^c(n_s/n_s^0 - 1), \tag{2.3}$$

$$j_p^v = j_0^v \cdot [1 - \exp(-q\Delta V_s/k_0 T)] = j_0^v(1 - p_s/p_s^0) \tag{2.4}$$

where j_0^c and j_0^v represent exchange currents characterizing the electron exchange between the electrolyte and the c and v bands of the semiconductor in equilibrium. In this case, the ion density in the electrolyte at a semiconductor surface is assumed to be large and not significantly different from the equilibrium one. These relations dictate the shape of the voltage-current curve and, correspondingly, the polarization characteristic. It is worth noting here that the latter

---

[2] These relations suggest that, under the application of an external voltage, the equilibrium distribution of free charge carriers is retained. Strictly speaking, this assumption holds only over limited ranges of currents and voltages. We discuss this issue at greater length below.

is usually determined by the exponential dependence of the surface density of electrons and holes in a semiconductor on the applied voltage, and, hence, is of the form

$$\varphi = a + b \log j,$$

which is known as Tafel's law [3].

Generally, the voltage-current relationship (the polarization characteristic) for the SE interface is represented as the dependency of the total current $j = (j_n^c + j_p^v)$ on the applied voltage, which suggests that both the semiconductor bands are involved in charge exchange. In the case of wide-band-gap semiconductors, however, due to a significant difference in the exchange currents for the bands (see Chapter 1), only one of the bands is usually involved in the electron exchange, namely, the one that is associated with bulk majority carriers. In this case the voltage-current curve features an asymmetric (rectifying) character, similar to the diode characteristic of a p-n junction [54, 55]. We note that, for n-type semiconductors, the forward direction corresponds to cathode polarization (downward band bending), whereas the reverse bias (back current) relates to anode polarization (the bands are bent upward). The reverse situation occurs in a p-type semiconductor.[3]

For narrow-gap semiconductors ($E_g < 1$ eV) one might expect that both c and v bands of the semiconductor will contribute, depending on the region of polarization, to the charge transfer across the interface. Here, the prevailing role of one of the bands is defined by both the direction and the amount of band bending as well as the potential drop in the Helmholtz layer, resulting in a change of the energy-level position of redox pairs with regard to the band edges on the semiconductor surface. In this case both consecutive and parallel contributions of the two bands to the charge transfer across the interface are possible. This situation is likely when the distribution functions of empty and occupied electronic states in the electrolyte overlap with both bands at once. This may take place for sufficiently narrow-gap semiconductors at certain relations between the value of the forbidden band and the Fermi level position in the semiconductor and the values of the redox potential and of the reorganization energy

---

[3] This analogy with the p-n junction is of a formal character, since there is an inherent difference between the origins of exchange currents in the case of a p-n junction and that of the SE interface.

9

in the electrolyte. Such an opportunity was probably first demonstrated on a germanium electrode of n type, where the injection of holes under cathode polarization (the effect of current multiplication) was revealed [38]. Hence, the general appearance of the voltage-current curve (polarization characteristic) may be quite intricate and defined by the sign and value of the applied voltage and also by the character of the bands involved in the electron exchange.

The features discussed above of the charge transfer across an interface under the application of an external field imply that the quasi-equilibrium distribution of free charge carriers within the SCL is preserved, a condition that is usually satisfied when majority carriers of a semiconductor are involved in the charge transfer. In fairly narrow-gap semiconductors, however, the charge transfer may occur also with the participation of minority (with respect to the semiconductor bulk) charge carriers. This situation can be realized with anode polarization for n-type semiconductors as well as with cathode polarization for p-type semiconductors. That may be accompanied by an appreciable disturbance of the quasi-equilibrium distribution of free carriers in the near-surface region of the semiconductor because of a lack of minority carriers in the semiconductor bulk and the finite velocity of their thermal generation and transport to the SE interface.

The role of minority carriers in charge transfer through a SE boundary was first indicated in [38], where germanium was investigated. It was shown that, for n-type samples in the anode region and those of p type in the cathode one, current growth with increasing applied voltage is restricted by the limiting flow of minority carriers, which is reflected in the occurrence of a saturation current in the polarization characteristics. Figure 2.1 shows representative examples of such a characteristic. A concurrent sharp increase of the electrode potential results from accumulation at the semiconductor surface of the electrolyte ions of relevant sign due to the fact that the conveyance of minority carriers required for their neutralization is limited. It is noteworthy that the saturation current grows with increasing specific resistance of Ge, which corresponds to an increase in the concentration of bulk minority carriers.

The saturation current value is defined as the limiting diffusion current flow of minority charge carriers from the quasi-neutral semiconductor bulk provided the condition of complete exhaustion of carriers at the SCL border with the quasi-neutral volume is satisfied.

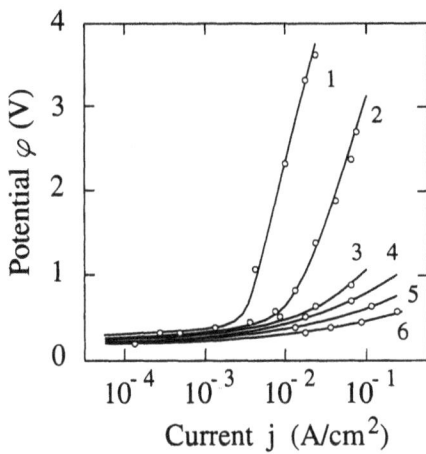

Figure 2.1 Polarization characteristics of Ge electrodes in 0.1 mol HCl aqueous solution: (1) n type, $10\,\Omega\cdot$cm; (2) n type, $6\,\Omega\cdot$cm; (3) n type, $25\,\Omega\cdot$cm; (4) p type, $30\,\Omega\cdot$cm; (5) p type, $6\,\Omega\cdot$cm; (6) p type, $1\,\Omega\cdot$cm.

The relevant relations for electrons and holes are, respectively, given in the form [54, 55]

$$j_n^{\lim} = qD_n(\mathrm{d}n/\mathrm{d}x) = qD_n(n_0/L_n) = qn_0\cdot\sqrt{D_n/\tau_n}, \qquad (2.5)$$

$$j_p^{\lim} = -qD_p(\mathrm{d}p/\mathrm{d}x) = -qD_p(p_0/L_p) = -qp_0\cdot\sqrt{D_p/\tau_p} \qquad (2.6)$$

where n_0 and p_0 are equilibrium electron and hole densities in the quasi-neutral volume of the semiconductor, D_n, D_p, L_n, L_p, τ_n and τ_p are the diffusion coefficients, diffusion lengths, and lifetimes for electrons and holes, respectively. In a more general case, these expressions, corrected for surface recombination and with current multiplication taken into account, are given in [3]. The role of minority charge carriers in the polarization characteristics is evidenced by the saturation current growth in the bulk under illumination, resulting in the additional generation of minority carriers, and also by the current decrease in the magnetic field, which is the result of the diffusion coefficient decrease.

A characteristic feature of polarization curves of such semiconductors as Si, GaAs, and others which have a fairly wide band gap ($E_g > 1$ eV) is a deviation from Tafel's law already at small polarization currents ($j \sim 1\,\mu A\cdot cm^{-2}$ for Si), in those cases when the current across the interface is defined by minority charge carriers. This is a result of their low density in the volume and their low thermal generation rate. And, as shown in Figure 2.2, there is no complete saturation of the

11

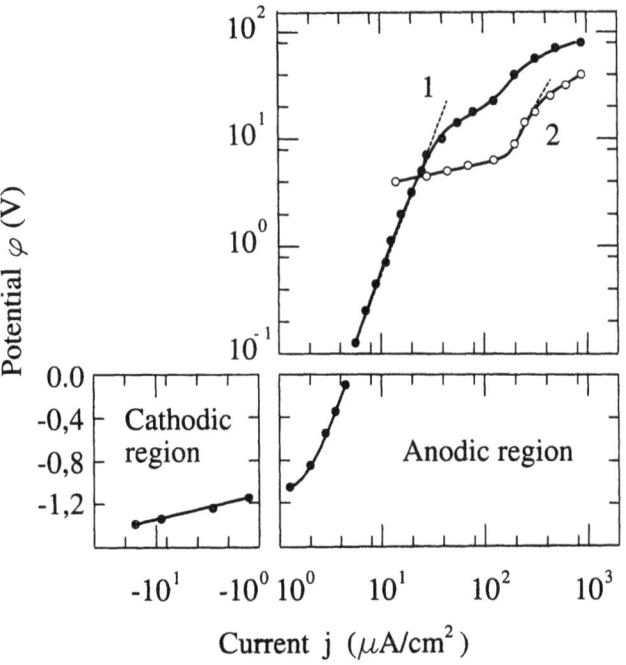

Figure 2.2 Polarization characteristics of Si electrode (n type, $\rho = 150\,\Omega \cdot$ cm) in 10 mol NaOH solution in darkness curve (1) and by light illumination curve (2).

current, which up to breakdown is quadratic in the electrode potential [56]. Such behavior of the polarization curves points to the fact that the number of minority charge carriers involved in the current through the interface is defined not by their diffusion from the quasi-neutral bulk but rather by their thermal generation rate in the near-surface region depleted of free carriers, whose thickness grows with increasing applied voltage up to the onset of the breakdown. There is experimental evidence showing that it is the small bulk density of minority carriers that accounts for the behavior of wide-band-gap semiconductors discussed above. This is the observation that, at fairly strong illumination of silicon, producing a sufficient increase of the bulk minority-carrier concentration, a complete restoration of the voltage-current characteristic typical for germanium is revealed. Consideration of electron transfer across the SE interface usually suggests that this process, when a so-called activated state is reached, takes place by means of tunneling of electrons across the barrier to the

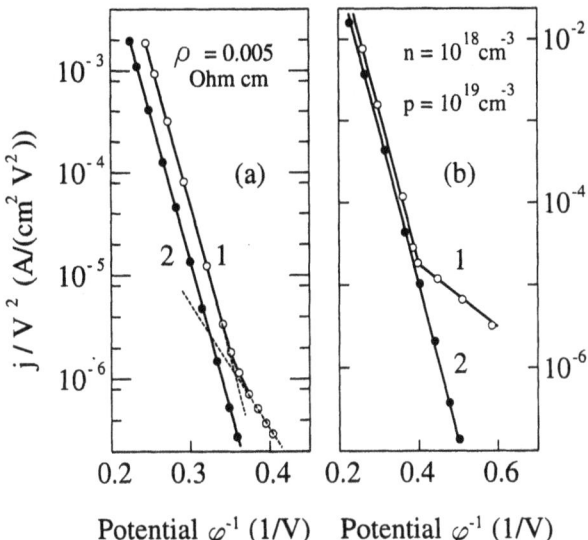

Figure 2.3 Current-voltage characteristics of degenerate Si (a) and GaAs (b) electrodes of *n* type (curves 1) and *p* type (curves 2) in 1 mol Na$_2$SO$_4$ solution.

phase boundary with a probability equal to unity [57] (see Chapter 1). With this assumption, the current value is independent of the barrier permeability. However, when a thin insulating film, transparent for electron tunneling, is present on the surface, there is likely a situation where the current across the SE interface depends on the changes in the permeability at the boundary with the applied voltage. This is especially true of semiconductors with degenerate electron and hole gases. In this case the voltage drop in the bulk and in the SCL is negligibly small; hence, all the major changes of the applied field occur in the insulating film on the semiconductor surface. Moreover, with the polarization direction along the flow of electrons from the semiconductor to the electrolyte (cathode polarization), the charge transfer process is similar, in a way, to the one occurring at field emission [58]. This statement is supported by the shape of the voltage-current curves, plotted, in Fowler-Nordheim coordinates, for silicon and gallium arsenide covered with a thin insulating film (Figs. 2.3a and 2.3b). For Si and GaAs of *n* type, these relationships represent two straight-line regions, corresponding to the emission of electrons from the *v* band (region 1) and *c* band (region 2). For samples of *p* type,

13

Figure 2.4 Dependence of the light emission intensity on the polarization current for n-GaAs electrodes with different degrees of doping in 1 mol Na_2SO_4 aqueous solution.

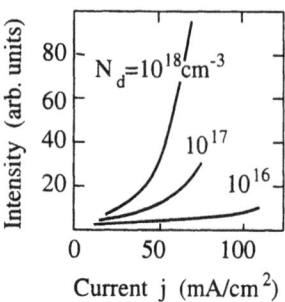

where the emission is only possible from the v band, the relevant relationships exhibit only one straight line with a particular slope corresponding to that characteristic of region 1 in n-type samples [59].

These considerations concerning the processes occurring are substantiated by the fact that the voltage-current characteristics are insensitive to illumination with light (absence of photosensitivity) for p-type samples and in region 1 for those of n type. Also, the above conclusions are corroborated by the fact that the relations between the slopes of these curves in regions 1 and 2 and the heights of the barriers, characteristic for the field emission of electrons from the v and c bands of Si, are in close agreement. As for GaAs the described character of the process is confirmed by the observation of electroluminescence in n-type samples with no emission revealed in those of p type. The occurrence of electroluminescence in n-type samples is due to the fact that the emission of electrons from the v band results in emptying the electronic states, to which transitions of electrons from the c band become feasible, yielding the emission of light. This explanation is corroborated by the increase of the emission intensity with the degree of doping and the polarization current (Fig. 2.4). The spectral distribution of the emission, where the main maximum observed in GaAs corresponds to band-to-band transitions, also provides support for the above explanations.

Quasi-Equilibrium Field Effect in Semiconductor-Electrolyte Interfaces: Studies of Surface States

THE POTENTIALITIES of the field effect in semiconductor-electrolyte interfaces (FESE) as applied to studies of semiconductor surface properties are based on the alteration of the semiconductor surface capacitance under polarization. This alteration reflects changes in the occupation of electron and hole states, with their energy levels distributed through the forbidden band (the fundamental bulk energy gap) as well as beyond it. The employment of the FESE also enables formation in the SE system of semiconductor surfaces with an extremely low density of surface states over the forbidden band, which is practically unattainable in other environmental conditions (vacuum, surrounding gas). This circumstance is helpful in studies of SCL characteristics. At the same time, it is of primary importance for the investigation of surface states arising in the course of electro-chemical and chemical reactions as well as on the adsorption from an electrolyte of various surface-active particles. In this case the initial surface with essentially no surface states may be used as a reference. To demonstrate the possibilities of the FESE technique, we shall discuss in the present chapter experiments on surface states arising on the germanium surface on chemical oxidation and adsorption of various different metals. Relevant experimental evidence on SCL characteristics obtained with the use of this technique is discussed in Chapter 5.

3.1   QUASI-EQUILIBRIUM FESE

Under polarization of a semiconductor in the electrolyte, a layer depleted of free charge carriers (depletion layer) or an accumulation layer arises in the region near the surface. In broad-band-gap semiconductors, the accumulation is usually realized through

15

majority carriers. In fairly narrow-gap ones, the accumulation can be accomplished by means of minority charge carriers, which corresponds to the formation of an inversion layer on the semiconductor surface. In that case, with the magnitude of the resulting current arising at polarization much below those corresponding to equilibrium exchange currents, the equilibrium distribution of free charge carriers is retained in the semiconductor surface region. Under these conditions the FESE may be characterized as a quasi-equilibrium FE in a semiconductor at the electrode boundary, which is similar to the FE in MIS structures. Strictly speaking, such an analogy is true for the cases when the experiment provides a wide enough range of electrode potentials related to the ideal polarizability of the semiconductor electrode. The ideal polarizability suggests that the conditions are satisfied when charge transfer through the phase boundary does not take place, that is, there are no electrochemical reactions at the SE interface. Note that in such a case the Helmholtz layer plays the role of an ideal insulator [52]. The range of electrode potentials corresponding to the ideal polarizability is a function of the electrolyte composition (in particular, of the attendant impurity content), the type of electrode, and the conditions of the experiment (temperature, stirring rate, etc.). Note also that, to a good approximation, one may consider that the polarization characteristics [$j(\varphi)$ curve], within the intervals of potentials and currents where the quasi-equilibrium of free carriers in a semiconductor is preserved, follow Tafel's law [3].

One of the essential advantages of the FESE with respect to the FE in MIS structures is that the SE-interface capacitance is largely contributed by the semiconductor surface capacitance. This enables realization of the so-called capacitance FE. The principal experimentally measured characteristic here is the semiconductor surface capacitance and its dependence on the electrode potential, changed by means of polarization of the semiconductor in the electrolyte. The total capacitance of the SE system can, in fact, be represented as that of capacitors connected in series, corresponding to the Helmholtz layer (C_H) and the semiconductor surface (C_s) capacitances. In contrast to a conventional electrostatic capacitor, the SE capacitance (C_{SE}) depends on the polarizing voltage and is defined as the differential capacitance of the system:

$$C_{SE} = dQ/d\varphi_0$$

where $Q = Q_{el}$ is the charge on the electrolytic plate of the SE capacitor; $\varphi_0 = (V_s + V_H)$ is the overall voltage drop across the interface, that is, the sum of the semiconductor surface potential (V_s) and the voltage drop in the Helmholtz layer (V_H). For the semiconductor plate, this charge is due to the surface charge $(Q_s = |Q_{el}|)$, which is made up of the charge in the SCL (Q_{sc}) and that in the surface states (Q_{ss}), namely, $Q_s = (Q_{sc} + Q_{ss})$. Then the capacitance here may be represented in the following form:

$$C_{SE} = \frac{dQ_{el}}{d\varphi_0} = \frac{dQ_s}{dV_s} \cdot \frac{dV_s}{d\varphi_0} = \left(\frac{dQ_{sc}}{dV_s} + \frac{dQ_{ss}}{dV_s} \right) \frac{dV_s}{d\varphi_0}. \tag{3.1}$$

The value of $dV_s/d\varphi_0$ is derived from the condition of the electric field continuity at the phase boundary and is given by

$$dV_s/d\varphi_0 = 1 - dV_H/d\varphi_0 = 1 - \frac{dQ \cdot dV_H}{d\varphi_0 \cdot dQ} = 1 - C_{SE}/C_H, \tag{3.2}$$

where $C_H = dQ/dV_H$ is the Helmholtz layer capacitance calculated per unit square; $dQ = dQ_{el} = |dQ_s|$. Substituting (3.2) into (3.1) we obtain the general expression for the differential capacitance of the semiconductor–electrolyte interface C associated with the capacitance of the SCL (C_{sc}) and the surface states (C_{ss}) of a semiconductor:

$$C_{SE} = (C_{sc} + C_{ss})(1 - C_{SE}/C_H) = \frac{(C_{sc} + C_{ss})C_H}{C_{sc} + C_{ss} + C_H}, \tag{3.3}$$

where $C_{sc} = dQ_{sc}/dV_s$ and $C_{ss} = dQ_{ss}/dV_s$. This capacitance can be represented by the equivalent circuit diagram as shown in Fig. 3.1. From theoretical and experimental estimates, the Helmholtz layer capacitance has the value of $C_H = 20\,\mu F/cm^2$ [3] which is usually well in excess of that corresponding to field capacitors on the basis

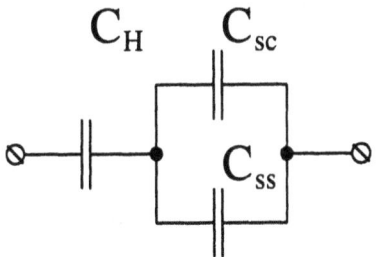

Figure 3.1 Equivalent circuit diagram of the semiconductor-electrolyte border.

of conventional MIS structures. As a result, the total experimentally measured capacitance of the SE system is

$$C_{SE} = \frac{(C_{sc} + C_{ss}) \cdot C_H}{(C_{sc} + C_{ss}) + C_H} = \frac{C_s \cdot C_H}{C_s + C_H} \approx C_s, \qquad C_s \ll C_H; \qquad (3.4)$$

in other words, it is mainly defined by the semiconductor surface capacitance. Physically, this is due to the fact that the main drop of the voltage applied to the SE interface occurs essentially within the SCL, in contrast to MIS structures, where it takes place in the insulator layer.

When the density of surface states on a semiconductor surface that has been brought into contact with an electrolyte is low enough, the measured capacitance of the SE interface is completely defined by that of the SCL. The theoretical dependence of the SCL capacitance on the surface potential in the absence of surface states and degeneracy is given by [8]

$$C_{sc} = \frac{dQ_{sc}}{dV_s} = \pm \frac{\varepsilon_0 \varepsilon_{sc}}{L_D \cdot F(u_b, v_s)} \cdot \frac{\sinh(u_b + v_s) - \sinh(u_b)}{\cosh(u_b)}, \qquad (3.5)$$

where $u_b = qU_b/(k_0 T)$ and $v_s = qV_s/(k_0 T)$ are dimensionless potentials; $U_b = (E_F - E_i)/q$;

$$L_D = \sqrt{\frac{\varepsilon_0 \varepsilon_{sc} k_0 T}{q^2 (n_b + p_b)}};$$

$$F(u_b, v_s) = \sqrt{2 \left(\frac{\cosh(u_b + v_s)}{\cosh(v_b)} - v_s \tanh(u_b) - 1 \right)}.$$

n_b and p_b are the equilibrium bulk electron and hole densities. For an intrinsic semiconductor ($n_b = p_b = n_i$) the dependence of C_{sc} on V_s represents a U-shaped curve with an exponential increase of the capacitance on both sides of the minimum. We note here that the magnitude of the capacitance at the minimum yields the value of the intrinsic carrier density in the semiconductor according to

$$n_i = \frac{k_0 T C_{min}^2}{2q^2 \varepsilon_0 \varepsilon_{sc}}. \qquad (3.6)$$

If we make a correlation between the experimental dependence of the capacitance on the electrode potential and the theoretical one $C_{sc}(V_s)$ by matching them at their common minimum ($V_s = 0$) it is possible to relate the scale of the electrode potential to that of the surface potential. For doped semiconductors, such an interrelation can be obtained by correlating the experimental capacitance-voltage characteristic with the theoretical one for a flat-band potential ($V_s = \varphi_{fb}$). Experimentally, the value of φ_{fb} is usually found by plotting the $C(\varphi)$ characteristic corresponding to the depletion layer in Schottky-Mott coordinates $[C^{-2}(\varphi)]$. In so doing, it is assumed that the measured capacitance is not influenced by the Helmholtz layer. A thorough inspection has demonstrated, however, that taking account of C_H changes the expression for the flat-band potential, which acquires the following form [17]:

$$\varphi_{fb}^* = \varphi_{fb} + \frac{k_0 T}{q} - \frac{\varepsilon_0 \varepsilon_{sc} q N_d}{2C_H^2}. \tag{3.7}$$

For heavily doped semiconductors, the last term in the formula may be appreciable in value. A representative example of capacitance-voltage curves for germanium electrodes of different types and degrees of doping is shown in Fig. 3.2. The sharp drop of the capacitance and the absence of a minimum in the capacitance-voltage curves for low-resistivity samples of p-type germanium in the anode

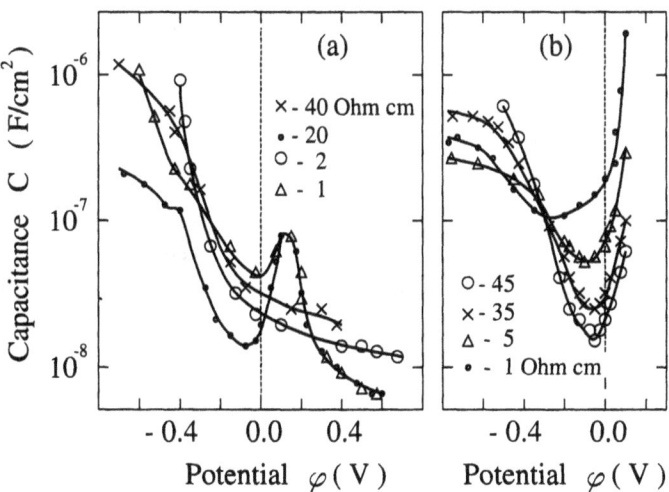

Figure 3.2 Capacitance-voltage characteristics of Ge electrodes of n type (a) and p type (b) in 0.1 mol Na_2SO_4 aqueous solution.

polarization region is due to the processes associated with non-equilibrium depletion [41, 60] (see Chapter 4).

With surface potentials corresponding to the onset of electron (hole) degeneracy in the semiconductor SCL, the dependence of the differential capacitance on the surface potential becomes less strong and transforms from exponential to power law [61]. If a semiconductor exhibits a square dispersion law for allowed bands, then $C_{sc} \sim V_s^{1/4}$. When we have nonsquare dispersion, say, of the Kane type, then $C_{sc} \sim V_s$. The semiconductor surface capacitance with surface states present incorporates their capacitance as well. In this case, the dependence of C_s on V_s should not correspond to $C_{sc}(V_s)$, and that dependence may not always be used for determination of the surface potential. We shall discuss this issue in what follows.

Another important characteristic measured in the FESE, which permits studies of SCL properties, is the dependence of the surface conductance on the surface and correspondingly on the electrode potential. The dependence of the surface conductance on the surface potential is defined, in the absence of degeneracy, by the relation

$$\sigma = q(\mu_{ns}\Delta N_s + \mu_{ps}\Delta P_s) \tag{3.8}$$

where μ_{ns} and μ_{ps} are the electron and hole mobilities at the semiconductor surface, respectively, and

$$\Delta N_s = \int_0^\infty [n(z) - n_b]dz = n_b L_D \int_{v_S}^0 \frac{\exp(+v) - 1}{\pm F(u_b, v)} \, dv, \tag{3.9}$$

$$\Delta P_s = \int_0^\infty [p(z) - p_b]dz = p_b L_D \int_0^{v_S} \frac{\exp(-v) - 1}{\pm F(u_b, v)} \, dv \tag{3.10}$$

are the excess surface electron and hole densities, respectively. The dependence of the surface conductance on the surface potential given by these relationships is represented by a U-shaped curve with exponential growth of the conductance on both sides of the minimum as V_s is increased. The position of the minimum corresponds to a value of V_s given by

$$V_s = -\frac{k_0 T}{q}\left(2u_b - \ln\frac{\mu_{ps}}{\mu_{ns}}\right). \tag{3.11}$$

With a significant amount of band bending (near the onset of the electron and hole degeneracy at the surface), the dependence of the

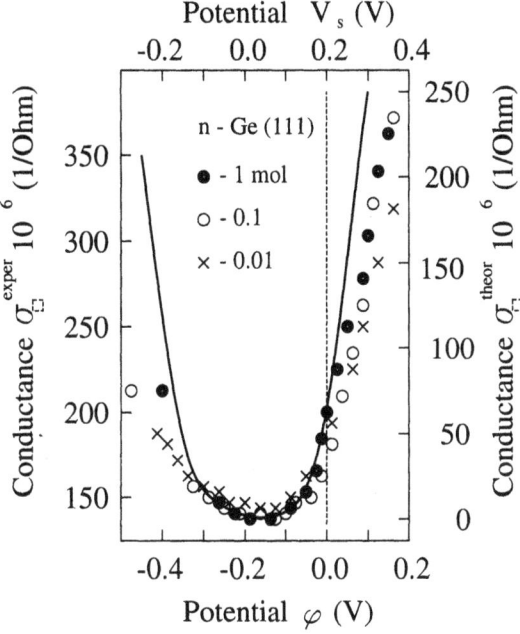

Figure 3.3 Experimental dependences of the conductance on the electrode potential for germanium (40 $\Omega \cdot$cm) obtained in the Na_2SO_4 aqueous solutions of different concentrations [42]. Solid line is the theoretical dependence of the surface conductance on the surface potential.

surface conductance on the surface potential becomes a power law, which is associated with the power-law character of changes of ΔN_s and ΔP_s [8].

Figure 3.3 presents the surface conductance as a function of the surface potential for germanium, assuming bulk mobility values. Also shown are the experimental dependences of the conductance on the electrode potential obtained in electrolytes of different concentrations; the relevant curves are brought into coincidence at their common minimum with the theoretical dependence of the surface conductance on the surface potential. A fairly good agreement between experiment and theory is found over a wide range of electrode potentials and makes possible a correlation between the electrode and surface potentials. In addition, this correlation shows that, even at the highest available electrolyte concentrations, when the electrolyte conductance well exceeds that of the bulk semiconductor, the electrolyte does not shunt the surface conductance of the semiconductor. This is useful in measurements allowing determination of the surface potential. These issues, and, in particular, the problem of shunting, will be discussed at greater length in Section 8.1.

The relevant formulas for surface conductance calculations usually suggest that the electron and hole mobilities are equal to their bulk values. Yet the employment of surface conductance measurements assuming bulk mobility values in studies of SCL properties may prove, in certain circumstances, to be incorrect because of a significant contribution to the mobility of surface scattering [62]. The latter is readily revealed in the case of a strong accumulation and inversion at the semiconductor surface, that is, under conditions when most of the free carriers determining the surface conductance are in the immediate vicinity of the surface. The correction for the mobility changes with varying electrode (surface) potential associated with the surface scattering may prove to be appreciable also near $V_s = 0$, since the power dependence defining the changes of the surface free-carrier density as a function of surface potential is weak in this region. Hence, to find a correct value of V_s (the relation between the surface and electrode potential scales), concurrent with studies of the conductance in the FESE, measurements of the surface mobility as a function of the electrode potential should be performed. The most popular techniques for mobility measurements employed in the FESE method are the Hall current and transverse magnetoresistance effects [63]. Note that the magnetoresistive mobility μ_{sm}, determined from the magnetoresistance coefficient ξ, coincides with the surface effective mobility:

$$\mu_{sm} = \frac{10^8}{H} \sqrt{\frac{\sigma_0 - \sigma_H}{\sigma_H}} = \xi \cdot \mu_{eff} \qquad \left(\frac{cm^2}{(V \cdot s)} \right) \qquad (3.12)$$

where σ_0 and σ_H are the semiconductor conductances in the absence of the transverse magnetic field H (Ga), and in its presence, respectively. The coefficient ξ is controlled by the surface scattering; relevant values of it for the various different mechanisms of scattering are listed in [63].

If one makes a comparison between measurements of the capacitance and conductance to determine V_s and characterize the SCL, the former are preferable, since there is no contribution to the capacitance from the free-carrier mobility. In addition, the dependence of σ on φ changes weakly near the minimum. The conductance

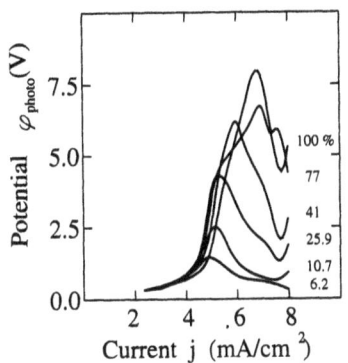

Figure 3.4 Photopotential of Ge electrode (n type, $\rho = 1\ \Omega \cdot cm$) as a function of the polarization current obtained in 0.1 mol Na_2SO_4 aqueous solution by different intensities of light illumination (in percent). 100% of light intensity corresponds to $15.2 \cdot 10^{-2}\ W \cdot cm^{-2}$.

may introduce tangible errors in determination of V_s as compared with the capacitance-voltage characteristics, especially at low enough values of V_s.[1]

However, with surface states present, the capacitance-voltage curves are essentially modified as compared to the theoretical relationship $C_{sc}(V_s)$, which makes it almost impossible to use them for determination of surface potentials. In this case, measurements of the surface conductivity should be used.

Another important characteristic of the FESE allowing the evaluation of surface potential changes is the photopotential, which represents the alteration of the electrode potential under illumination of a semiconductor. This is exemplified by Fig. 3.4 where the photopotential $\Delta\varphi_{photo}$ as a function of the polarization current for n-type germanium is presented. These results were obtained for anode polarization in Na_2SO_4 solution. As seen from the figure, the curves resemble the polarization characteristics (see Chapter 2); they exhibit a Tafel region, a current saturation region, and a breakdown region. The values of $\Delta\varphi_{photo}$ relating to the saturation of the photopotential dependence on the illumination intensity enable determination of the amount of surface band bending, namely, V_s.

In addition to studies of SCL properties, the FESE provides opportunities for investigating the electronic states on a semiconductor surface. That is because, within the total measured semiconductor capacitance $C_s = C_{sc} + C_{ss}$, the constituent capacitances C_{sc} and C_{ss} can be separated because of their different dependences on surface

---

[1] For the case of the germanium electrode changes in the value of V_s by 100–150 meV lead to only a few percent change in the surface conductance [8].

and electrode potentials, respectively. This makes possible investigations of such characteristics of surface states as their density and energy distribution through the semiconductor forbidden gap. In the presence of an open semiconductor surface there exists the possibility of controlled change of its properties and monitoring of the effects on its physical and chemical characteristics of various external perturbations. This offers wide possibilities for studying the origin of surface states and their behavior, which is one of the crucial problems in semiconductor surface physics.

In conducting such studies, analytical expressions for the dependence $C_{ss}(V_s)$ for various types of energy distributions of surface states in the forbidden band are used. When surface states are characterized by their energy levels, we have for the capacitance the following relation:

$$C_{ss} = \frac{q^2 N_{ss}}{4 k_0 T} \left[\cosh\left(\frac{E_{ss} - E_i}{k_0 T} \right) - (u_b + v_s) \right]^{-2}, \qquad (3.13)$$

where N_{ss} and E_{ss} are the density and energy of the surface states [8]. This expression represents a curve exhibiting a maximum, which arises in the course of recharging of surface states under polarization, on the background curve depicting the variation of the SCL capacitance. Note that the position of maximum capacitance on the V_s scale assesses the energy position of the surface state level in the semiconductor forbidden band, whereas the magnitude of the capacitance relating to the maximum gives the density of surface states[2]: $N_{ss} = (4 k_0 T / q^2) C_{max}$. For a uniform distribution of surface states over energy in the semiconductor forbidden band, we have

$$C_{ss} = \frac{q^2 N_{ss}}{4 k_0 T},$$

while for an exponential distribution

$$C_{ss} = \frac{\pi \alpha}{\sin(\pi \alpha)} \cdot \frac{q^2 \mathcal{A}}{k_0 T} \cdot \exp(\alpha \mathcal{E}_{ss}),$$

where α and $\mathcal{A}$ are the parameters of the exponential distribution $N_{ss} = \mathcal{A} \cdot \exp(\alpha \mathcal{E}_{ss})$; $\mathcal{E}_{ss} = E_{ss}/(k_0 T)$. Most frequently, the energy distribution of surface states in the gap exhibits a U-like quasi-continuous

---

[2] This relation can be generalized to include surface states having several discrete energy levels [8].

character [64, 65]. In this case, the expression for the dependence of the capacitance on the surface potential is of the form

$$C_{ss} = \mathcal{K} \sqrt{\varepsilon_0 \varepsilon_{sc} \frac{(\eta/\pi)^{5/8}}{(4 a_n a_p)^{3/8}} \cdot \exp(-\varepsilon_g/4\delta)}, \qquad (3.14)$$

$$\mathcal{K} = \frac{\cosh\dfrac{(v_s - v_0)}{2\delta}}{\sqrt{\sinh\dfrac{(v_s - v_0)}{2\delta}}}, \qquad \varepsilon_g = E_g/(k_0 T),$$

where a_n and a_p are the Bohr radii of electron and hole, respectively; δ characterizes the energy scale of charge fluctuation on the surface; $v_0 = [\varepsilon_g/2 + (3\delta/4) \cdot \ln(m_n/m_p)]$; and η characterizes the capture cross section of the corresponding surface state [64].

Another means of surface state investigations is the one based on the measurements of surface recombination velocity (S) as a function of the surface (electrode) potential [8]. In the case of discrete energy levels of surface states these measurements allow one to obtain the energy position of the surface center as well as to assess the relation between the capture cross sections of electrons and holes by that center. Due to the symmetrical shape of the $S(V_s)$ curve, the position of the level associated with the surface state is not unambiguous; in fact, it is of a two-valued character. Hence, to obtain the value of E_{ss}, measurements are carried out at various different temperatures. A generalization of $S(V_s)$ to the case of a quasi-continuous distribution of surface states over the band gap was carried out in [66]. Obtaining $S(\varphi)$ and correspondingly the $S(V_s)$ dependence in a SE system is based on measurements of either stationary photoconductance or photocapacitance. In the latter case we have

$$S \approx C_{dark}^2/(C_{photo}^2 - C_{dark}^2), \qquad (3.15)$$

where C_{dark} and C_{photo} are the values of capacitance measured in the dark and under semiconductor surface illumination, respectively.

3.2   SURFACE STATES INDUCED BY GERMANIUM SURFACE OXIDATION

We made use of the chemical oxidation of germanium (Ge) in aqueous solutions of HNO_3 which is an essential constituent of a number

25

of etchants used to cleanse surfaces by oxidation followed by the removal of the oxides by dissolution. It is known that at a HNO_3 concentration of about 5 M, the oxidation speed of germanium dissolution shows a maximum, relating to the equality between the rates of oxidation and removal of the oxide [67, 68]. As a result, the oxide layer is practically absent and the surface becomes most clean and homogeneous. Such a surface has a minimal surface state density, which is attested by the low values of the surface recombination velocity ($S \sim 25 \, \text{cm/s}$). As the HNO_3 concentration is increased, an oxidic phase is observable on the germanium surface, which, as follows from electron-microscopic and electronographic evidence, is initially formed as separated islands corresponding to hexagonal GeO_2. Formation of an inhomogeneous surface is revealed in the abrupt increase in surface recombination velocity, a fact that indicates an increasing density of surface states. With further increase of HNO_3 concentration, a continuous uniform coating is produced. This brings about the decrease of the surface state density, which corresponds to the diminishing of the surface recombination velocity. The dependence of C on φ, measured for different HNO_3 concentrations and transformed into $C(V_s)$ curves, is depicted in Fig. 3.5.

Figure 3.5 Capacitance-voltage characteristics of Ge electrode (n type, $\rho = 40 \, \Omega \cdot \text{cm}$) depending upon dilution of HNO_3 with water.

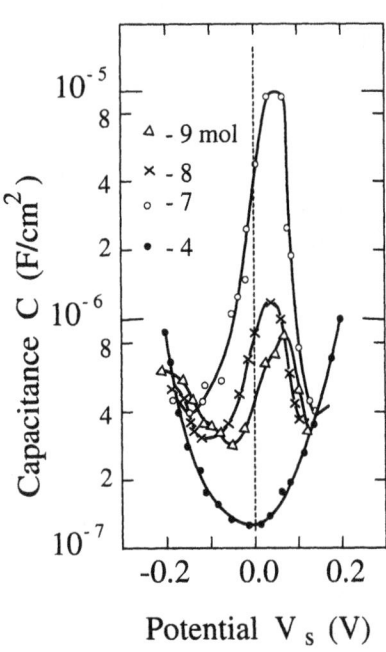

The curves demonstrate a modification of the $C(\varphi)$ dependence as one proceeds from the homogeneous oxide-free germanium surface [the U-like $C(\varphi)$ dependence coincident with the theoretical $C_{sc}(V_s)$ relationship] to that where the oxidic phase is present [69]. The latter is initially revealed in the form of separate islands with subsequent continuous uniform coating. As this takes place, the relationship $C(\varphi)$ acquires the form of a curve exhibiting a maximum, whose magnitude is diminished as the oxidic phase is formed. The results indicate that the occurrence of surface states at oxidation is largely due to inhomogeneities of the surface because of the inhomogeneous character of the oxidic phase formation. The surface states produced in the course of oxidation can be characterized on the basis of the shape of the $C(\varphi)$ relationship, by the discrete energy level that lies $(2-4) \cdot k_0 T/q$ above the mid-gap. The maximum density of these states is $N_{ss} = 5, 6 \cdot 10^{12}$ cm^{-2} and is lowered as a continuous coating is formed with the energy-level positions unchanged.

3.3 Surface States Arising on Germanium by Metal Adsorption

The possibility of FESE employment for surface state studies on the germanium surface on metal adsorption was first demonstrated by Brattain and Boddy [70]. Romanov, Konorov, and co-workers [42, 71–73] have performed a series of experimental investigations of surface states arising due to the adsorption of various different metals. In these studies the initial germanium surface was obtained by means of chemical and subsequent anode etching in aqueous solutions of Na_2SO_4. For such a surface, the relevant capacitance-voltage $C(\varphi)$ characteristics followed the theoretical relationship $C_{sc}(V_s)$. That pointed to a low density of surface states, which was not in excess of 10^9 cm^{-2} near the midgap. Metal ions of different concentrations were introduced in the Na_2SO_4 solution in the form of the relevant metal salts with subsequent measurements of the electrode potential and FESE characteristics.

These investigations have demonstrated that all the metals explored can be classified in relation to their effect on surface properties into two groups situated on the right and on the left of germanium in the succession of standard potentials of oxidoreduction reactions. The metals situated on the right of germanium (the so-called electropositive metals) produce pronounced changes of electrophysical

Figure 3.6 Capacitance-voltage curves of Ge electrode (n type, $\rho = 40\,\Omega\cdot\text{cm}$) obtained in 0.1 mol Na_2SO_4 aqueous solution before (curve 1) and after (curve 2) Cu^{2+} ion incorporation into electrolyte.

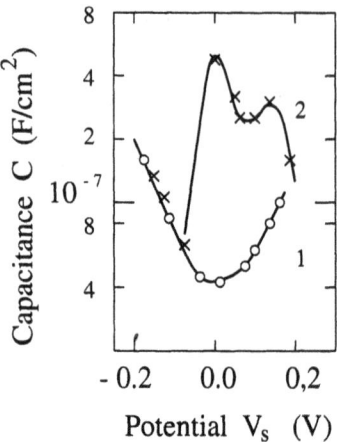

properties of the germanium surface. Their adsorption is accompanied by the occurrence of surface states whose energy levels lie near the midgap of germanium. The adsorption for electronegative metals did not bring about significant alterations of the surface and was not efficient in producing a substantial number of surface states.[3]

Figure 3.6 shows representative examples of $C(\varphi)$ characteristics after their transformation into $C(V_s)$ curves obtained in Na_2SO_4 solution before and after incorporation of Cu^{2+} ions. The latter curve exhibits two maxima which can be interpreted as resulting from the occurrence of surface states, having two discrete levels lying at -20 and $40\,\text{meV}$ with respect to the middle of the forbidden band of germanium. The position of one of these levels relates to that of the Cu atom, it being the bulk impurity center in germanium. The occurrence of two maxima in the capacitance-voltage curve is also seen in the adsorption of a number of electropositive metals.

Since the observation of two discrete surface state levels is typical for metals featuring increased ability to diffuse into Ge [74], it is suggested that the occurrence of one of these levels results from the penetration of metal atoms into the near-surface region enriched with vacancies. The position of the other level when thermal smearing is taken into account is practically the same for all electropositive

---

[3] The surface activity of electronegative metals appeared following subsequent cathode polarization which is associated, according to Romanov and co-workers [71], with the discharge of the adsorbed metal ions and their conversion to the atomic state.

metals. Its occurrence may be explained as resulting from distortions of the homogeneous surface which appear because of the precipitation of a metal in the atomic state, giving rise to single clusters at the surface. These assumptions are supported by the fact that there is no direct correlation between the density of surface states and the metal concentration in the solution. Note that their density is 3–4 orders of magnitude less than the total number of metal atoms isolated on the germanium surface. What is more, the density correlates with the number of metal clusters on the surface. In the case of Cu adsorption, these statements are supported by the following facts: first, the level appearing as a result of Cu adsorption and associated with the "bulk" Cu level correlates with the level of copper atoms in germanium vacancies [74]; second, the heights of both maxima in the capacitance-voltage curve differ in their behavior on anode dissolution of the germanium surface layer. The maximum that corresponds to the "surface" level is first removed from the spectrum.

In conclusion we note that in this chapter the potentialities of the FESE for studies of surface states were demonstrated. Those were exemplified by investigations performed on germanium for the case when the region of ideal polarizability is confined within a fairly narrow interval of surface potentials (a few tenths of eV) at about the midgap. In Chapters 5 and 6 we shall discuss methods allowing the extension of the ideal polarizability region to offer further possibilities for the employment of the quasi-equilibrium FESE technique.

Field Effect in Semiconductor-Electrolyte Interfaces on Nonequilibrium Depletion of Free Charge Carriers from Semiconductor

A GENERAL EQUILIBRIUM condition for a semiconductor SCL, manifested in that the Fermi quasi-levels must be equal throughout the specimen, is a balance between the drift and diffusion flows [66]. For a SE interface, this condition corresponds to a small value of the resulting current across the interfacial boundary with reference to its drift and diffusion components [3]. The balance is disturbed when charge transfer across the boundary occurs with participation of minority carriers in conditions when their flow to the surface is limited. This corresponds to the case when the saturation current is approached in the polarization characteristics. In this case, the semiconductor near-surface region becomes depleted of free charge carriers, which affects the FESE characteristics. An essential condition for observation of nonequilibrium depletion is a low enough rate of minority-carrier generation in the SCL region, as well as on the semiconductor surface itself, the latter being in contact with the electrolyte.

From the point of view of the FESE, two stages of nonequilibrium depletion may be distinguished. The first one arises as a result of minority-free-carrier exhaustion in the so-called quasi-neutral region of the semiconductor, the thickness of which is dictated by the diffusion displacement length of minority carriers. In this case a near-surface inversion layer may remain, and its thickness is determined by the Debye screening length L_D. The boundary between the quasi-neutral region and the Debye layer represents the surface; the diffusion current flow due to minority carriers through this equals the drift current. This situation is spoken of as diffusion depletion, since

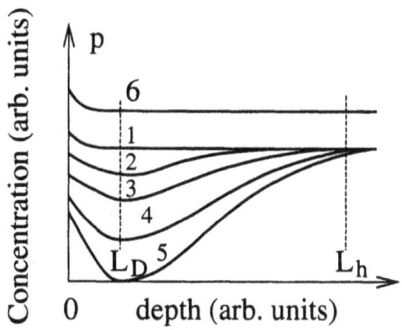

Figure 4.1 Schematic diagram of the spatial distributions of minority carriers (holes in n-Ge) near the semiconductor electrode surface by different anodic currents: (1) quasiequilibrium distribution; (2–5) distributions corresponding to increasing anodic current; (6) distribution by illumination.

the minority-carrier flow is governed by diffusion while the drift can be neglected. The associated spatial distributions of minority-carrier density are schematically represented in Fig. 4.1.

The second stage of nonequilibrium depletion, which is achieved at a certain field strength at the interfacial boundary, results from the disintegration of the inversion layer near the semiconductor surface. The double-layer electric field is spread deep into the interior of the semiconductor, accompanied by exhaustion of both minority and majority charge carriers. At the first stage, that is, diffusion depletion, screening of the semiconductor bulk occurs due to the inversion layer charge. At the second one the screening is accomplished through ionized donors and acceptors in the near-surface region, which is completely depleted of free charge carriers. The thickness of this layer is controlled by the density of completely ionized impurity centers and grows with increasing voltage up to the onset of the breakdown.[1] This type of depletion is referred to as field depletion, as distinguished from diffusion depletion. It is evident that diffusion depletion should essentially affect the characteristics of narrow-gap semiconductors with a low content of impurities and a fairly high concentration of minority charge carriers. In wide-gap semiconductors, the relevant characteristics are largely affected by nonequilibrium field depletion, which may take place without the diffusion stage. Yet in the SE system field depletion may exist permanently even in fairly narrow-gap semiconductors, such as Ge and Si, because of the absence of an inversion layer screening the semiconductor volume. In this respect the SE

---

[1] Within the framework of the present book, we do not discuss possible mechanisms of breakdown. For narrow-gap semiconductors ($E_g < 1$ eV) the most likely is that due to the Zinner effect, suggesting an electrostatic ionization mechanism. For wide-gap semiconductors, avalanche carrier multiplication takes place [54, 55].

system differs from MIS structures, in that it is only realized in the pulsed mode [75].

The existence of diffusion depletion was directly demonstrated using an optical method which makes use of IR emission and absorption by free charge carriers [76]. This technique enables one to detect the changes in free-carrier concentration in the course of depletion; it also permits study of their density distribution throughout the specimen. Measurements using a modulated probing IR beam passing through a high-resistivity Ge sample of n type showed that under anode polarization in aqueous electrolytes, the response is sharply enhanced with increased polarization current and is saturated at a current density corresponding to the limiting diffusion flow of minority charge carriers [77]. A quantitative estimate shows that there is a complete depletion of free carriers (holes) in germanium in the vicinity of the electrolyte boundary. Studies of the efficiency of the IR modulation as a function of the position of the probe beam revealed that it decays exponentially as the probe beam is shifted deeper into the semiconductor from the SE boundary; the decay constant corresponds to the relevant diffusion displacement of holes in Ge.

The diffusion character of the nonequilibrium depletion was also substantiated by conductance and Hall effect measurements [60]. The results obtained on high-resistivity Ge specimens of n type are presented in Fig. 4.2.

A drop in the conductance and a sharp increase in the value of the Hall emf with increasing polarization current are in qualitative and quantitative agreement with the diffusion depletion of minority charge carriers in the specimen. In this case a close-to-intrinsic Ge specimen with bipolar conductivity is converted into a monopolar semiconductor with a low electron conductivity. Also remarkable is the high light sensitivity of the Hall emf in depletion conditions; it is several times as much as the light sensitivity of the conductance of the same sample under the same conditions.

The occurrence of a region of complete exhaustion of free carriers as a consequence of field depletion in semiconductors at the SE boundary was, probably, first observed by Williams [78, 79] on CdS (see Fig. 4.3). Later, this effect was observed on Si, GaAs, and some other wide-gap semiconductors [56, 80]. The effect of the nonequilibrium field depletion as well as its influence on the FESE characteristics was also observed on low-resistivity n-type Ge specimens in the

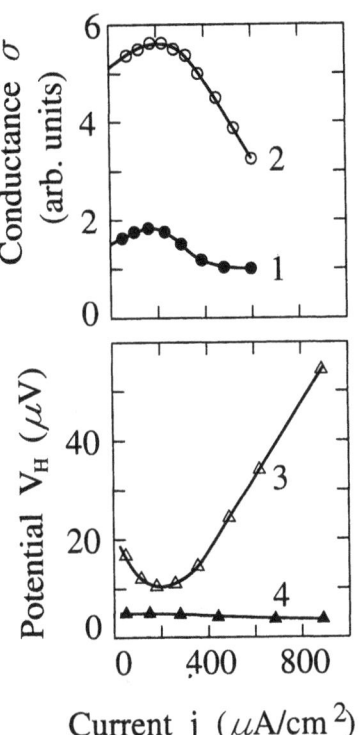

Figure 4.2 Variations in conductance (curves 1 and 2) and Hall effect (curves 3 and 4) with the anodic current density for the electrode (n type, $\rho = 40\ \Omega \cdot$cm). Curves 1 and 3 in darkness; 2 and 4 under illumination.

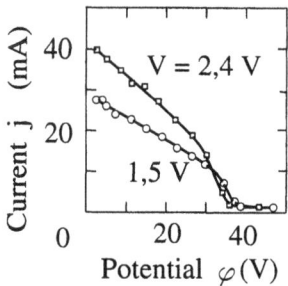

Figure 4.3 Variation in the current along the CdS crystal plate with the electrode potential for two different values of voltage, applied to the ends of the plate.

anode polarization region, where it is revealed as an abrupt drop of the capacitance (see Fig. 3.2) and conductance (see Fig. 4.4). For thin high-resistivity specimens, both stages characteristic of depletion, namely, diffusion and field stages, can be observed to transform into one another successively with increasing polarization current [60].

In conclusion, we point out some characteristic features of semiconductor behavior in the nonequilibrium state which are associated

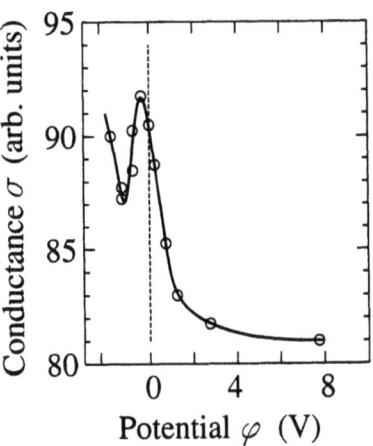

Figure 4.4 Conductance of the germanium electrode (*n* type, $\rho = 10\,\Omega \cdot cm$) as a function of the electrode potential in 0.1 mol Na_2SO_4 aqueous solution in the anodic polarization region.

with a high sensitivity of their parameters to any of the perturbations yielding an increase in minority-carrier density. First, these features manifest themselves as significant changes of the capacitance, conductance, and potential of the semiconductor under illumination from within the intrinsic absorption region. Another important factor that brings about high photosensitivity is a strong decrease in the efficiency of surface recombination due to spatial separation of the illumination-induced electron-hole pairs. The separation results from the electric field localized in the depletion layer. As a consequence, the surface recombination channel is "switched off," which results in a photosensitivity increase, especially noticeable in the short-wavelength region [81]. This effect is illustrated in Fig. 4.5, where the photoconductivity spectra are presented for a Si sample with both surfaces in contact with an electrolyte and different in the value of the applied polarization voltage. On illuminating the surface at which the nonequilibrium depletion region is built up, the value of the photoconductivity exceeds by more than an order of magnitude that arising under illumination of the opposite surface, where quasi-equilibrium is retained. In addition, there appears a maximum in the short-wavelength spectral region which is several times as much as the one at the band edge, which is characteristic of equilibrium conditions [82].

Among other perturbations capable of influencing the surface generation of minority charge carriers, deformation of the specimen (e.g., bending) should be mentioned. Figure 4.6 shows a reversible

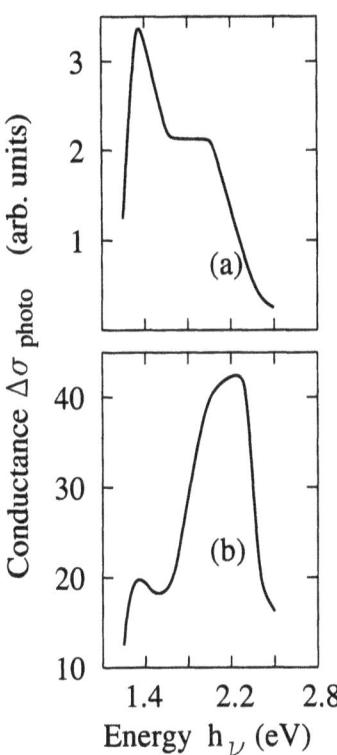

Figure 4.5 Spectral distributions of the photoconductivity for the Si electrode in 0.1 mol Na_2SO_4 solution: illuminated surface is (a) in quasiequilibrium state; (b) in nonequilibrium depletion state.

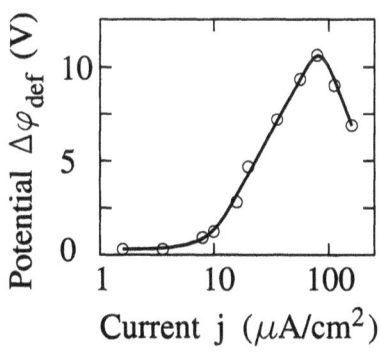

Figure 4.6 Variation in the electrode potential of a thin n-Ge plate ($\rho = 40\ \Omega\cdot cm$) with the anodic polarization current measured by its bending.

change in the electrode potential of a thin Ge plate by bending, which may be as high as a few volts.

The occurrence of a nonequilibrium depletion layer at a semiconductor surface in an electrolyte, along with the optical transparency of the latter, favors electroreflectance measurements, widely used for investigations of semiconductor band structure [83].

35

Application of Field Effect in Semiconductor-Electrolyte Interfaces for Studies of Surface Charge Layer Characteristics of Semiconductors and Semimetals

$\mathbf{M}$AJOR PROBLEMS in studying surface electrophysical properties of most binary and compound semiconductors are associated with the complexity of production of MIS structures with a high-quality insulator on their surfaces and a defect-free interphase boundary [62, 84, 85]. Because of this, starting from the mid-1970s, the SE system has been favored in detailed studies of surface electrophysical properties of these materials as well as in production of high-performance oxides (mostly anodic) on their surfaces [21, 85–87]. This especially applies to narrow- and zero-gap semiconductors and semimetals. These materials are of interest not only from the point of view of their practical applications (employment in high-speed electronics and optoelectronics), but also for studies of fundamental properties of solids whose band structure is radically modified at large Z in the periodic table. It is well known that relativistic effects play an important role in the formation of the band structure in the semiconductors with large Z. The dispersion law in such materials changes, as E_g is decreased, from parabolic to Kane type, and, in the limit of zero-gap semiconductors and semimetals, features ultrarelativistic behavior [88].

Application of the FESE first permitted investigations of electrophysical characteristics of surfaces of a large number of binary and ternary compounds on the basis of A^3B^5, A^2B^6, and A^4B^6 semiconductors. This technique makes it possible to study semiconductor surfaces in a wide range of surface potentials, including the region of band bendings corresponding to electron and hole degeneracy at the surface. This method can also yield information on many semiconductor band parameters at room temperature, which is often impossible to obtain with conventional techniques. The method is simple and does not require the preparation of metal-insulator-semiconductor structures and/or the use of low-temperature measurements, which makes it rather promising as applied to the nondestructive rapid determination of semiconductor surface and subsurface layer parameters.

In the present chapter we consider some results of these studies carried out on wide-, narrow- and zero-gap semiconductors and also semimetals.

5.1 WIDE-GAP SEMICONDUCTORS

5.1.1 GaAs

The properties of the GaAs surface brought into contact with an electrolyte are characterized mainly by the features of the natural surface. Investigations of the surface state density distribution on a p-type semiconductor surface revealed two maxima within the forbidden band with a total density of 10^{13} cm^{-2} [89–92]. The same distribution is characteristic of natural surfaces of GaAs of n and p types; it corresponds to the donor-acceptor model for defects, treated in [93]. The amount of band bending on the GaAs surface is on average about 0.4 eV and weakly dependent on the character of the redox pairs in the electrolyte [94]. This value of band bending corresponds to the Fermi level pinning on the GaAs surface characteristic of a natural surface coated with a thin layer of the intrinsic oxide [89, 95].

The capacitance-voltage characteristics and the impedance of the GaAs-electrolyte system have been studied by a number of researchers [95–102]. The characteristics exhibited, most commonly, a Schottky-Mott behavior, typical for depletion of the SCL near the surface of the semiconductor electrode. Figure 5.1 shows capacitance-voltage and current-voltage curves for electrodes of n (Figs. 5.1a and 5.1b) and p type (Figs. 5.1c and 5.1d), taken in the potentiodynamic regime. The doping impurity density derived from the Schottky-Mott relationships was found to be in close agreement with the specifications of the given samples. A characteristic feature of the presented dependences for the samples of n type was the constant capacitance in the region of cathode potentials corresponding to the accumulation of electrons on the GaAs surface. Similar characteristics were also obtained for the case when the etching perixo-ammoniacal solution

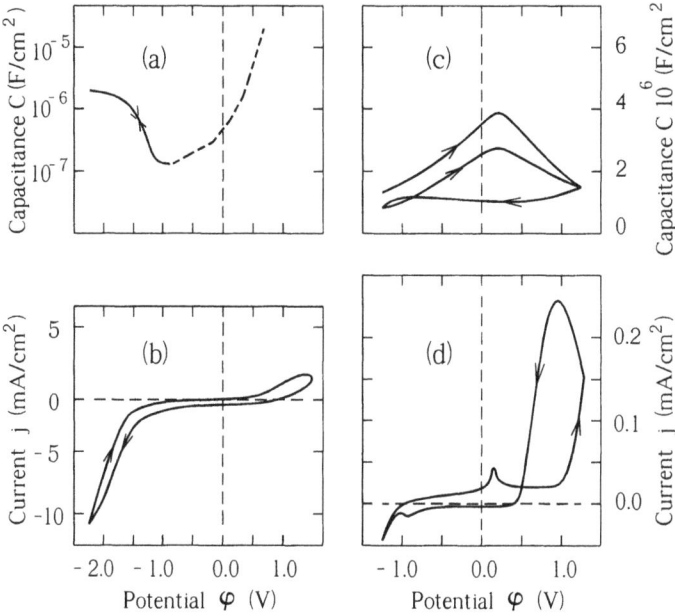

Figure 5.1 Capacitance-voltage (a), (c) and current-voltage (b), (d) characteristics of n-GaAs (a), (b) and p-GaAs (c), (d) electrodes in 1 mol Na_2SO_4 aqueous solution (pH = 5–6) (dotted parts of the curves are not reproduced on cycling).

was used as the electrolyte. The absence of dispersion of capacitance-voltage characteristics was found within 1–200 kHz, which points to the absence of surface states with relaxation times of 10^{-3}–10^{-4} s in the range of surface potential variation. The values of the constant capacitance were found to be well below the Helmholtz layer capacitance.

The results of investigations of the system n-GaAs–electrolyte are presented in Figure 5.2.

We note that the constancy of the capacitance in the capacitance-voltage curve cannot be accounted for by the presence of oxide on the GaAs electrode surface, since it is revealed in the etching solution and its occurrence is always accompanied by an increase of the current through the interphase boundary. This contradiction is explained by the fact that the constancy of the capacitance with varying electrode potential is the result of Fermi level pinning on the surface [62], which is likely to be due either to the high density of charge in surface states

Figure 5.2 Capacitance-voltage (a) and current-voltage (b) characteristics of n-GaAs electrode in 1 mol KCl + EDTA aqueous solution (pH = 8).

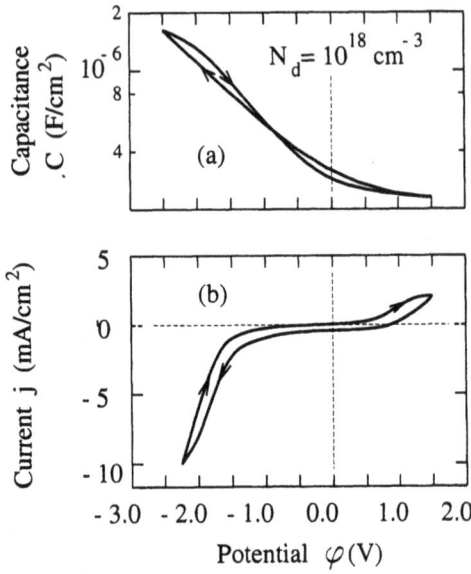

with relaxation times comparable with the probe signal, or to the onset of the corrosion process in this potential region.

A characteristic feature in this case is the realization of a wide range of surface potentials, including, along with a depletion region, strong accumulation of electrons on the GaAs surface. This indicates the possibilities of electrolytic contact for reduction of the effect of Fermi level pinning on the FE characteristics.

5.1.2 GaP and GaPAs

Surface properties of GaP electrodes were studied in [100, 103–105]. Representative capacitance-voltage and voltage-current curves for aqueous electrolytes are shown in Figure 5.3 [103, 104]. In order to dissolve the hydroxide compounds arising on the GaP surface under polarization, a substance that initiates complex formation was first introduced into the electrolyte. With $\varphi > 0$ in the anode polarization region, the capacitance-voltage relations have a shape typical of the capacitance-surface potential relations for the case of depletion of the SCL; they flatten in Schottky-Mott coordinates.

The energy scheme of the GaP–aqueous electrolyte contact is represented in Figure 5.4. Coincidence of the Fermi level in GaP with nE_{dec} is seen at $\varphi = -0.54$ V. At the same potential value, capacitance saturation in the $C(\varphi)$ dependence is revealed (see Fig. 5.3a).

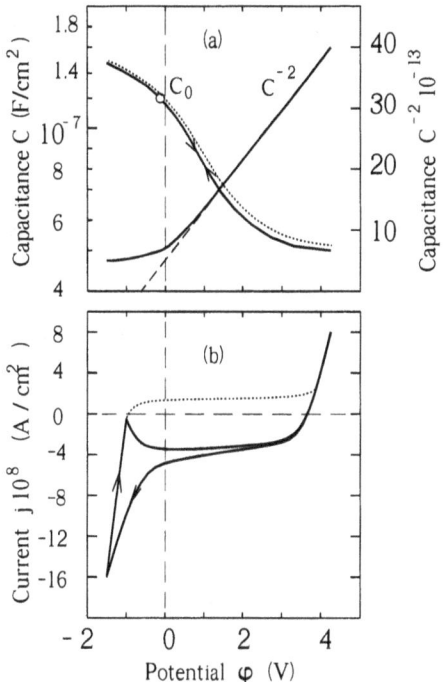

Figure 5.3 Capacitance-voltage (a) and current-voltage (b) characteristics of the n-GaP ($N_d = 1.8 \cdot 10^{17}$ cm^{-3})–electrolyte (KCl + EDTA) system (dotted line, under light illumination).

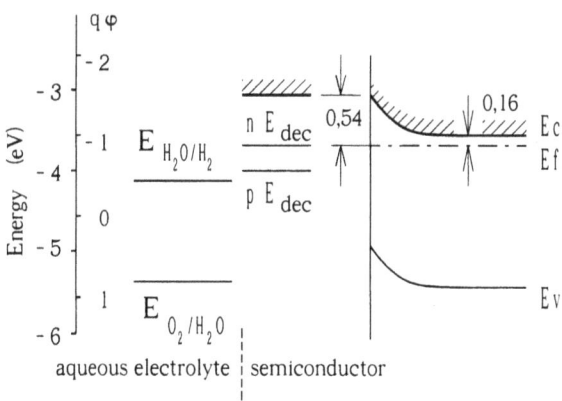

Figure 5.4 Schematic energy diagram of the n-GaP aqueous electrolyte contact; nE_{dec} and pE_{dec} are the energies of electrochemical decomposition connected with employment of the electrons and holes of the electrode, respectively.

41

This means that the saturation of the capacitance accompanied by an increase of the current in the voltage-current characteristics (see Fig. 5.3b) is due to the fact that the electrochemical reaction of GaP dissolution takes place rather than the formation of an oxide layer or pinning of the Fermi level due to surface states.

Similar characteristics were obtained using the FESE applied to solid solutions of GaP_xAs_{1-x}, starting from GaP, with a thin layer of intrinsic oxide on the surface [104]. These findings allow us to characterize the GaP_xAs_{1-x}–intrinsic oxide–electrolyte system for all values of x within $0 \leqslant x \leqslant 1$ as a liquid Schottky diode. Here, forward and reverse biases correspond to cathode and anode polarizations, respectively.

5.1.3   GaN

Electrochemical and photoelectrochemical properties of GaN were investigated in [106, 107].

The results of FESE studies for GaN electrodes in aqueous electrolyte are presented in Figure 5.5 [43, 108]. The GaN specimens were epitaxial large-block films doped with Zn with concentrations of $(1-5) \cdot 10^{17}$ and $(0.5-1.0) \cdot 10^{15}\,cm^{-3}$. The current-voltage characteristics have the form of charging curves complicated by some peculiarities caused by Zn doping of GaN (Fig. 5.5a). The capacitance-voltage curves (Fig. 5.5b), obtained with the relaxation time dispersion in the FESE pulse taken into account, acquire a straight-line shape in Schottky-Mott coordinates. The density of the doping impurity in GaN was found to be in correspondence with the specifications of the material used. The density distribution for surface states in the forbidden band of GaN obtained from these data is illustrated in Figure 5.6 (curve 1). A comparison between the surface state density distribution and the spectral distribution of light emission from diodes formed from GaN crystals doped with Zn [109] (curve 2) points to their similar behavior within the energy range from 1.9 to 2.8 eV. The qualitative agreement between the surface state and light emission distributions allows one to ascribe the features observed in the surface state distribution in GaN epitaxial films to the Zn impurity precipitating on the surface. The observed high values of the conductance (Fig. 5.5c) are likely to be conditioned by the bulk conductivity shunting due to the impurity band appearing under doping of GaN with Zn, yielding a set of donor levels with the energies $E_d - E_v = 1.7, 2.2, 2.6,$ and $2.8\,eV$ [109]. The Fermi level lies between the levels

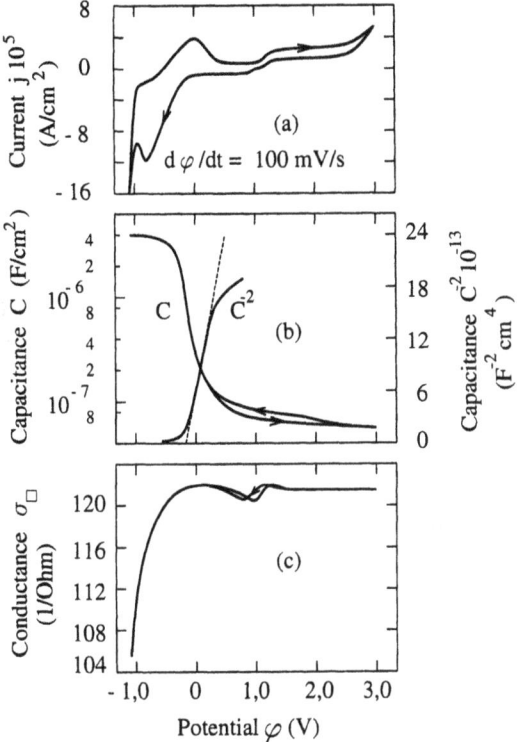

Figure 5.5 Polarizing current (a), capacitance (b), and surface conductivity (c) as functions of the electrode potential for GaN–0.1 mol KOH system.

$E_d - E_v = 2.6$ and 2.8 eV. The occupation of the latter level on the GaN surface will govern its surface conductance.

5.1.4 InP

The results for surface properties of InP electrode investigations are presented in [87, 100, 103, 110–112].

Studies in aqueous electrolytes, including heavily concentrated acidic and basic solutions, showed [110, 113] that at minimal oxide layer thicknesses $[d_{ox} < (10–15)\,\text{Å}]$ capacitance-voltage and voltage-current curves, taken in the dark for n-type electrodes, demonstrate, independent of the electrolyte pH, characteristics typical of a liquid Schottky diode, forward biased in the cathode polarization region and reverse biased in the anode polarization region (Fig. 5.7). In all cases the reverse dark current in the polarization characteristics

Figure 5.6 Surface state density and light emission energy distributions for GaN.

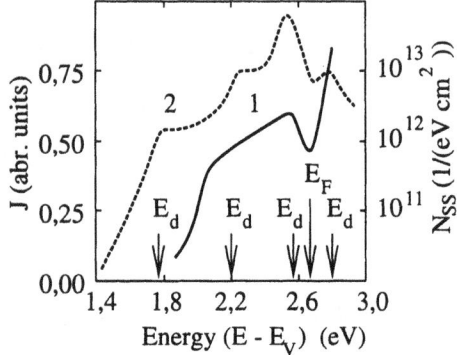

Figure 5.7 Capacitance-voltage (a) and current-voltage (b) characteristics of the n-type InP electrode, surface A^3-(111), in 0.1 mol H_2SO_4 (curve 1), 0.1 mol Na_2SO_4 (2), and 0.1 mol NaOH (3).

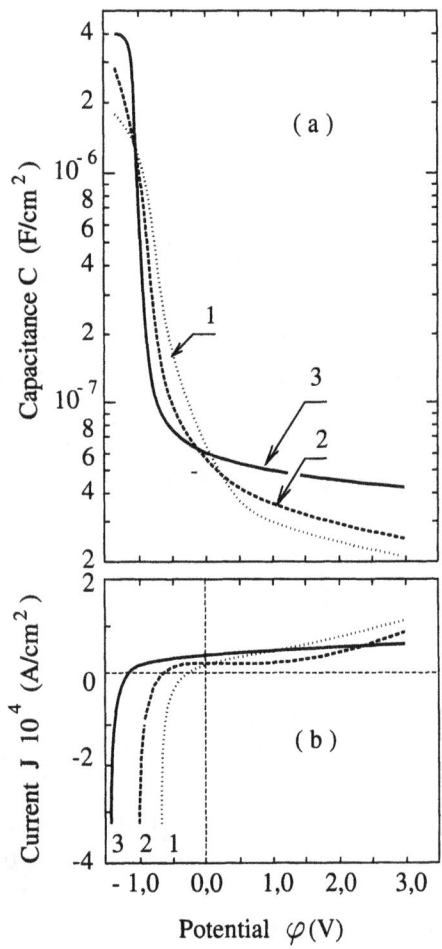

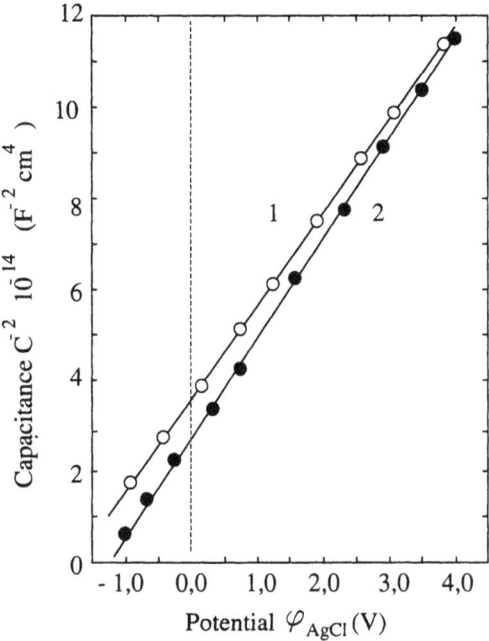

Figure 5.8 Capacitance-voltage curves in Schottky-Mott coordinates for InP electrode (n-type, $N_d = 4.5 \cdot 10^{16}$ cm^{-3}) in 0.1 mol NaOH: (1) surface A^3 (111); (2) surface B^5 (111).

remains within 0.5–1 mA/cm^2, which greatly exceeds the theoretically calculated limiting diffusion current due to minority charge carriers ($j_s \sim 10^{-16}$–10^{-18} A/cm^2) and indicates a substantial contribution to the reverse current of generation processes in the InP SCL and on the InP-electrolyte boundary. Dark capacitance-voltage curves reflect, for the most part, changes of the semiconductor SCL capacitance in the region of accumulation (cathode region) and nonequilibrium depletion (anode region). In the latter case, the curves are in close agreement with the calculated ones for a semiconductor SCL (the Schottky-Mott law) (Fig. 5.8) [110].

5.2 Narrow-Gap Semiconductors

Narrow-gap semiconductors at room temperature usually represent intrinsic semiconductors with charge carrier density exceeding

10^{16} cm^{-3}. Even at a small amount of band bending they are characterized by degeneracy of the electron (hole) gas near the surface. This allows us to exploit the potentialities of the FESE for studying a number of band parameters in the near-surface region, namely,

- the energy distribution of effective electron and hole state densities;
- the character of the dispersion law;
- the absolute values of effective masses and state densities for electrons and holes;
- the forbidden band width, etc.

Consider in some detail the technique by which these parameters can be obtained within a zero-temperature approximation [61, 114]. In the latter, thermal smearing is neglected; this approximation is fairly good at room temperature for strong degeneracy, when the Fermi level on the semiconductor surface falls into the allowed band at a distance $\approx 2k_0 T$ from its edge. In this case, the charge alteration in the SCL of a semiconductor following a change of the surface potential by dV is given by the relation d$Q = \rho(V_s)$dx, where $\rho(V_s)$ is the charge density on the surface, and dx is the change of the SCL width with the surface potential change.

Then, taking into account that

$$C_{sc} = -\frac{dQ}{dV_s} = -\rho(V_s)\frac{dx}{dV_s}, \qquad (5.1)$$

the Poisson equation can be represented in the following form:

$$\frac{d}{dV_s}\left(\frac{\rho(V_s)}{C_{sc}(V_s)}\right) = -\frac{C_{sc}(V_s)}{\varepsilon_0 \varepsilon_{sc}}. \qquad (5.2)$$

Since d$E = q$dV_s and within the zero-temperature approximation for $\rho(V_s) = -qN(E)$, the equation (5.1) can be written in the form

$$\frac{d}{dE}\left(\frac{N(E)}{C_{sc}(V_s)}\right) = -\frac{C_{sc}(V_s)}{q^2 \varepsilon_0 \varepsilon_{sc}}. \qquad (5.3)$$

Upon integrating the equation (5.3) one can obtain the following expression for the differential density of states in the allowed band:

$$\frac{dN}{dE} = \frac{1}{q^2 \varepsilon_0 \varepsilon_{sc}}\left(\frac{dC_{sc}(V_s)}{dV_s}\int_{V_{s1}}^{V_{s2}} C(\varphi)dV + C_{sc}^2(V_s)\right). \qquad (5.4)$$

In this expression, the density of electronic states corresponds to the energy $E = q(V_s - V_z)$, where $V_z = (E_c - E_F)/q$ reckoned from the c band edge and depends on the Fermi level position in the semiconductor volume.

The interval of integration in (5.4) is chosen in accordance with the investigated region of the allowed c or v band, respectively.

The relation (5.4) allows determination from capacitance-voltage relation measurements in the region of strong degeneracy of the distribution of differential density of electron (hole) states, that is, dN/dE as a function of E, and hence characterization of the dispersion law in the allowed bands.

Thus, for a quadratic dispersion law, one has

$$N(E) = \frac{16\pi\sqrt{2}}{3h^3} m_{de}^{3/2} E^{3/2}, \qquad \frac{dN}{dE} = \frac{8\pi\sqrt{2}}{h^3} m_{de}^{3/2} E^{1/2}, \qquad (5.5)$$

and therefore

$$C_{sc} = g\varepsilon_{sc}^{1/2} \left(\frac{m_{de}}{m_0}\right)^{3/4} (V_s - V_z)^{1/4}, \qquad (5.6)$$

where

$$g = \left(\frac{900\pi^2}{g} \frac{m_0^3 q^5 \varepsilon_0^2}{h^3}\right)^{1/4} = 8.979 \cdot 10^{-6} \left(\frac{F}{cm^2} V^{1/4}\right).$$

For the Kane dispersion law

$$E(p) = \frac{E_g}{2}\left[\left(1 + \frac{2p^2}{E_g m_{de}}\right)^{1/2} - 1\right] \qquad (5.7)$$

and the expression for the state density is of the form

$$N(E) = \frac{8\pi}{3h^3}(2m_{de}E_g)^{3/2}\left[\left(\frac{E}{E_g}\right)^2 + \left(\frac{E}{E_g}\right)\right]^{3/2}, \qquad (5.8)$$

$$\frac{dN}{dE} = \frac{8\pi\sqrt{2}}{h^3} m_{de}^{3/2} E_g^{1/2}\left[\left(\frac{E}{E_g}\right)^2 + \frac{E}{E_g}\right]^{1/2}\left(\frac{2E}{E_g} + 1\right). \qquad (5.9)$$

47

The accurate solution of equation (5.4) is somewhat cumbersome. Yet for $E < E_g/3$, the c band, to a good approximation, may be considered as parabolic. At $E \geq E_g/2$ the following expression holds true:

$$C_{sc} = g \left(\frac{8}{5} \right)^{1/2} \varepsilon_{sc}^{1/2} \left(\frac{m_{de} q}{m_0 E_g} \right)^{3/4} \left(|V_s - V_z| + \frac{E_g}{2q} \right). \qquad (5.10)$$

It follows from (5.6) and (5.10) that the slopes of $C_{sc}(V_s)$ curves allow one to determine the values of $\varepsilon_{sc}^2 m_{de}^3$ and $\varepsilon_{sc}^{1/2}(m_{de}/E_g)^{3/4}$ for quadratic- and Kane-type dispersion laws, respectively.

5.2.1   GaSb

Current-voltage curves and capacitance-voltage characteristics of n- and p-type GaSb electrodes taken in 0.1 mol aqueous KNO_3 electrolyte for various crystallographic orientations of the surface are represented in Figures 5.9–5.11. These characteristics feature the regions of electrode potentials corresponding to accumulation, depletion, and inversion. In the absence of polarization, the

Figure 5.9 Capacitance-voltage (a) and current-voltage (b) characteristics of n-GaSb electrode [surface A^3 (111), $N_d = 5 \cdot 10^{17}$ cm^{-3}] in 1 mol KNO_3 aqueous solution. Dotted line shows the influence of illumination.

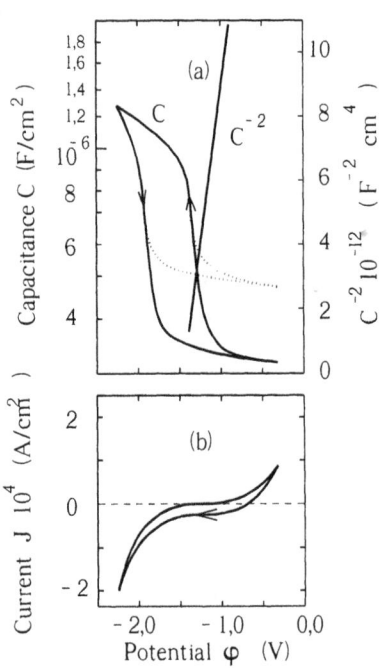

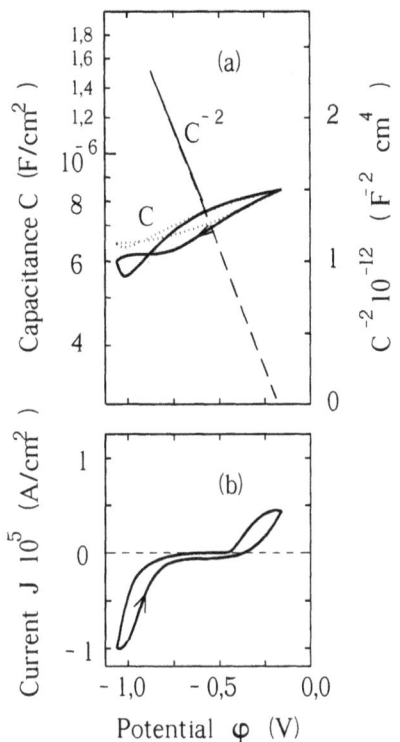

Figure 5.10 Capacitance-voltage (a) and current-voltage (b) characteristics of p-GaSb electrode [surface A^3 (111), $N_d = 2 \cdot 10^{18}$ cm^{-3}] in 1 mol KNO$_3$ aqueous solution. Dotted line shows the influence of illumination.

surface potential values are 0.4–0.5 V for n-GaSb and 0.2 for p-GaSb, which correspond in each case to the depletion of the surface by majority charge carriers. These values are not functions of the electrolyte composition and, for different surface orientations, obey the inequality $V_{so}^{(111)} < V_{so}^{(100)} < 0$. Such values of V_s are, evidently, the result of Fermi level pinning by donor and acceptor surface states induced by vacancy-type intrinsic defects on the GaSb surface, which are of inherent origin. A change of the electrolyte pH value by introducing HCl and NaOH in the range of pH from 6 to 11 shifts φ_{fb} toward lower values (in the absolute value scale) of the cathode potential (see Fig. 5.12) accompanied by a coincidental current density polarization increase.

Measurements of the field effect on the p-GaSb electrode surface performed within a wide range of electrode potentials show [108] that two regions of capacitance variation can be distinguished in the electronic branch of the capacitance-voltage characteristic: the electrode potential region, where $C \sim \varphi^{1/4}$, and the region where fine

Figure 5.11 Capacitance-voltage (a) and current-voltage (b) characteristics of n-GaSb ($N_d = 5 \cdot 10^{17}$ cm^{-3})–1 mol KNO$_3$ aqueous solution system with different surface orientations. C_0 is the capacitance without polarization.

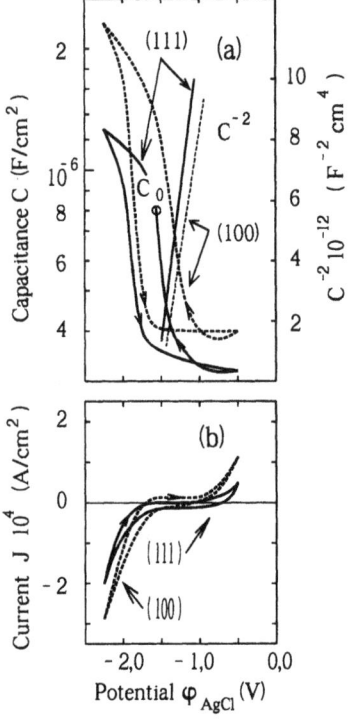

Figure 5.12 Variation in the flat-band potential with electrolyte pH for n-GaSb electrode.

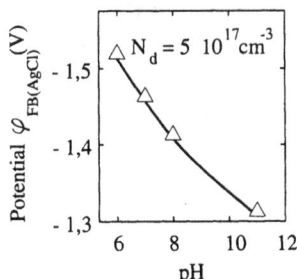

structure in the $C(\varphi)$ curve arises (see Fig. 5.13). The occurrence of the first region points to strong degeneracy of electrons on the gallium antimonide surface. Measurements of the $C(\varphi)$ characteristic in this region enable the effective mass of the electronic states density to be estimated in the conduction band of GaSb. That was found to be $m_{de} = 0.06m_0$ and close to the bulk value of m_{de} in gallium antimonide crystals.

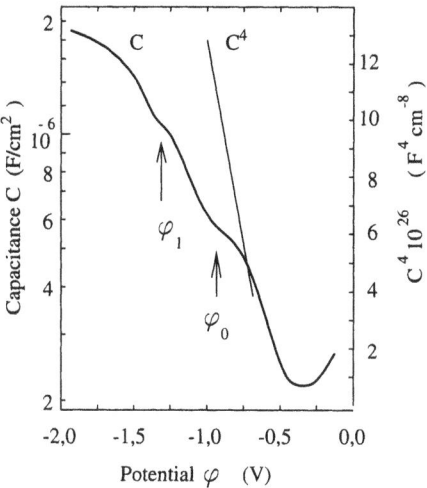

Figure 5.13 Variation in the capacitance with the electrode potential for p-GaSb ($N_a = 2 \cdot 10^{18}$ cm^{-3}) electrode in aqueous electrolyte with EDTA (φ_1 and φ_0, see Section 7.2.)

5.2.2 InAs

The capacitance-voltage dependences for polar A^3 and B^5 facets of n-InAs for different electrolyte pH values have a minimum, which is typical of narrow-gap semiconductors (Fig. 5.14) [115]. The capacitance minimum in acidic and alkaline solutions is located higher than the theoretical minimum calculated for the SCL of the semiconductor, unlike that in neutral solutions. This minimum is shifted, respectively, toward the regions of anodic and cathodic potentials. Similar behavior of these dependences with solution pH variation is typical for all narrow-gap semiconductors, including Ge and GaSb.

The tendency of the capacitance to saturation in the cathodic region of electrode potentials results from the existence, on the InAs surface, of an ultrathin intrinsic oxide layer, the thickness of which does not exceed 50–90 Å [115]. Analysis of these curves shows that the differential capacitance in the region of cathodic polarization is proportional to the surface potential. This type of dependence is characteristic for a degenerate electron gas at a surface with a nonparabolic (Kane-type) dispersion law. This enables determination of the value of m_{de}/E_g. With $\varepsilon_{sc} = 11.7$ and $E_g = 0.35$ eV, the value of the effective mass of the electrons in the conduction band of InAs was found to be $m_{de} = 0.014 m_0$ which is somewhat below the bulk value of m_{de} and may be due to an error brought about by the neglect of the ultrathin oxide layer capacitance and that of the Helmholtz layer.

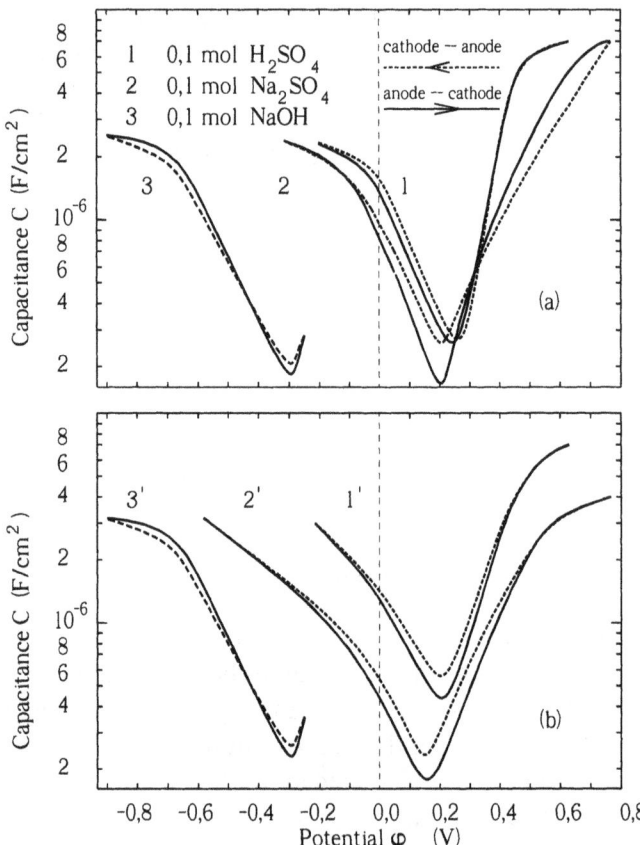

Figure 5.14 Capacitance-voltage characteristics of n-InAs electrode ($N_d = 2 \cdot 10^{16}$ cm^{-3}) in electrolytes with different pH: (a) surface A^3 (111); (b) surface B^5 (111).

Figure 5.15 demonstrates the distributions of the differential density of electron states obtained from the $C(\varphi)$ characteristics (curve 1) and calculated on the assumption of a quadratic (curve 2) and a Kane-type dispersion law (curve 3) of the conduction band. It follows from the figure that the conduction band in the near-surface region of indium arsenide more correctly follows the Kane model.

The dependence of the magnetoresistive mobility of electrons in the inversion channel of p-InAs on the electrode potential is obtained from surface conductance measurements with and without a magnetic field in the region of quadratic dependence of $\Delta\sigma_0/\sigma_0$ on the magnetic field strength (Fig. 5.16). This dependence is bell-like,

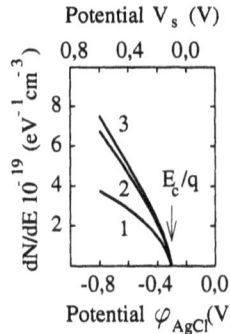

Figure 5.15 Distributions of the differential density of electron states in n-InAs, obtained from the experimental $C(\varphi)$ characteristic (curve 1) and calculated for different dispersion laws (curves 2 and 3).

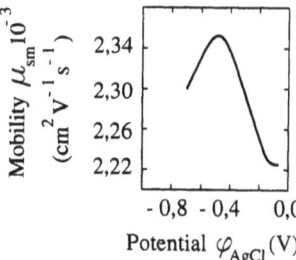

Figure 5.16 Dependence of the magnetoresistive mobility of electrons on the electrode potential for p-InAs $[N_a = 2 \cdot 10^{14}$ cm^{-3}, surface $A^3(111)]$ in 1 mol Na$_2$SO$_4$.

a shape associated with scattering processes. The mobility values are in good agreement with the data known from the literature on measurements of electron mobility on indium arsenide surfaces at room temperatures [108].

5.2.3 InSb

FESE measurements showed that, in aqueous electrolyte solutions in the absence of polarization, the InSb surface is always characterized by the storage of a positive charge, independent of the electrolyte pH. Note that on indium antimonide there is almost always present an ultrathin layer of intrinsic oxide with a thickness of $d_{ox} < 20$ Å [103, 114]. The capacitance-voltage characteristics have features typical of an intrinsic semiconductor (Fig. 5.17). One can distinguish regions of electrode potentials corresponding to the accumulation of electrons and holes on the InSb surface and a minimum in the curve related to the flat-band state. The analysis of $C(\varphi)$ characteristics of the system InSb – 1 mol Na$_2$SO$_4$ measured with the electrode potential given potentiodynamically has shown that at $\varphi < -0.8$ V, on the

Figure 5.17 Capacitance-voltage characteristics of the *i*-InSb electrode in 1 mol Na_2SO_4 aqueous solution (curve 1) compared with curves (2 and 3) calculated at $T = 0$ and 300 K.

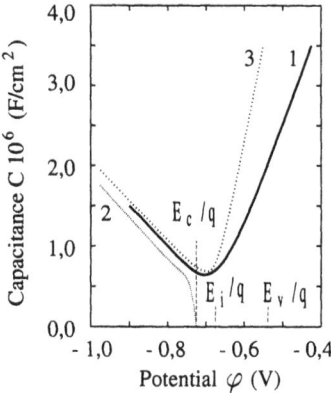

Figure 5.18 Capacitance-voltage curves for the *i*-InSb superthin anodic oxide film–electrolyte system.

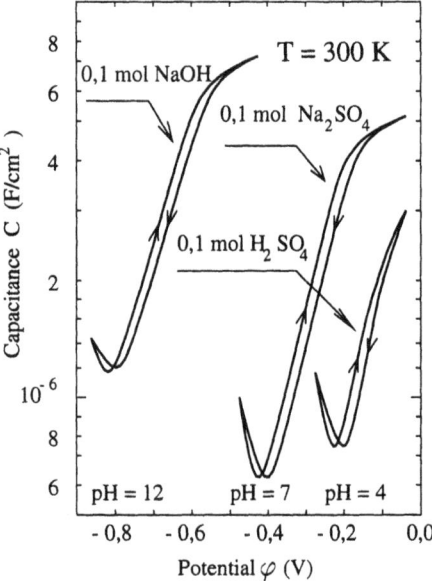

surface of InSb, the electron gas is degenerate and described by the Kane dispersion law [114]. This is evidenced by the linear character of the capacitance dependence on the potential. An estimate of the effective mass of the electronic states density in the conduction band on the InSb surface yields a value of $0.013 m_0$, in good agreement with its bulk value for the conduction band in InSb single crystals [116].

The dependence of $C(\varphi)$ curves on the electrolyte pH value is presented in Figure 5.18 [108]. These dependences concide qualitatively

54

with the corresponding dependences for Ge and InAs (see Figs. 6.4 and 5.14).

5.2.4 PbTe, PbSe, and PbS

The general appearance of the capacitance-voltage and current-voltage characteristics in the ideal polarizability region and the results of their analysis are given in [117–120]. Typically, PbTe (Fig. 5.19) and PbS (Fig. 5.20) electrodes show, when unpolarized, a p-type conductivity on the surface. The capacitance value at the minimum of the $C(\varphi)$ curve allows estimation of the intrinsic concentration of free charge carriers; the results are in good agreement with the data from independent Hall effect measurements. The analysis of $C(\varphi)$ relationships in the potential regions of electron and hole degeneracy at the surface shows that the dispersion law for allowed bands in the near-surface volume region is parabolic. The extrapolations of the relevant $C(\varphi)$ dependences up to the intersection with the potential axis (Figs. 5.19a and 5.20a) yield, therewith, values for the forbidden gap width in these materials that agree closely with the bulk results. The parabolic dispersion law for allowed bands at the surface of these materials is substantiated by measurements of the temperature dependence of the minimal capacitance [120].

Capacitance measurements in the degenerate region yielded effective masses of electrons and holes in the relevant allowed bands in

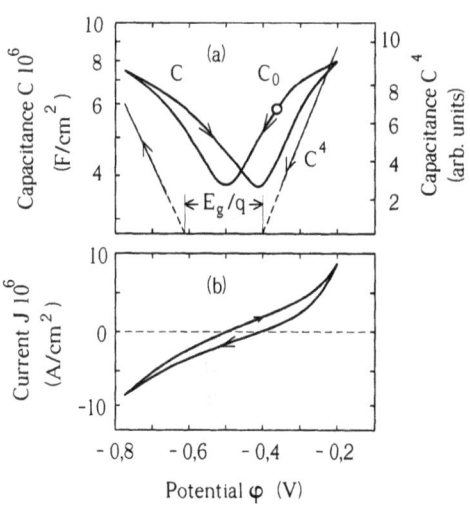

Figure 5.19 Capacitance-voltage (a) and current-voltage (b) characteristics of i-PbTe electrode in 1 mol $KNO_3 + KOH + EDTA$ aqueous solution (pH = 11).

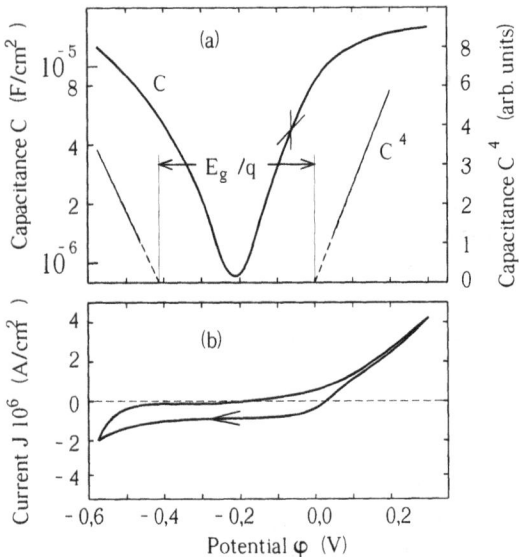

Figure 5.20 Capacitance-voltage (a) and current-voltage (b) characteristics of i-PbS electrode in 1 mol KCl aqueous solution.

the near-surface volume layer of PbTe and PbS which were found to be nearly equal to the bulk values. The dispersion law for the allowed bands near the surface of these materials can be described satisfactorily in the parabolic model [121, 122].

Capacitance-voltage and current-voltage characteristics of the system PbSe-electrolyte (Fig. 5.21) have a more complicated shape than those for PbTe and PbS and cannot be adequately interpreted within a simple equivalent scheme for the SE boundary.

5.2.5   TlBiSe$_2$

Ordered TlBiSe$_2$ layers were grown by the Bridgman-Stockbarger method [123–125]. Measurements were taken on complementary surfaces obtained from one crystal by delamination on cleavage planes. For realization of field-effect methods a saturated aqueous solution of KCl was used as the electrolyte. $C(\varphi)$ characteristics were measured with a potentiostatic cyclic change of electrode potential at a rate of change 10–100 mV/s, at room temperature $T = 300$ K.

The experimental $C(\varphi)$ characteristics look like the characteristics of n-type semiconductors (see Fig. 5.22). There is hysteresis and a

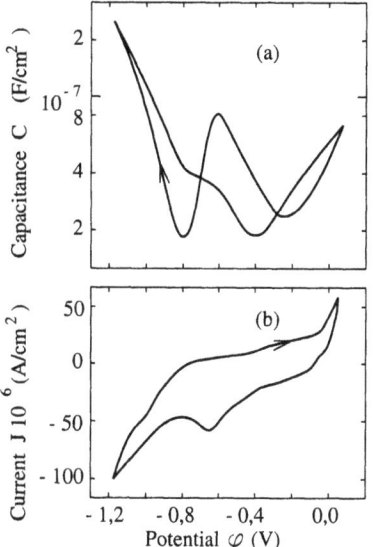

Figure 5.21 Capacitance-voltage (a) and current-voltage (b) characteristics of i-PbSe electrode in 1 mol KCl aqueous solution.

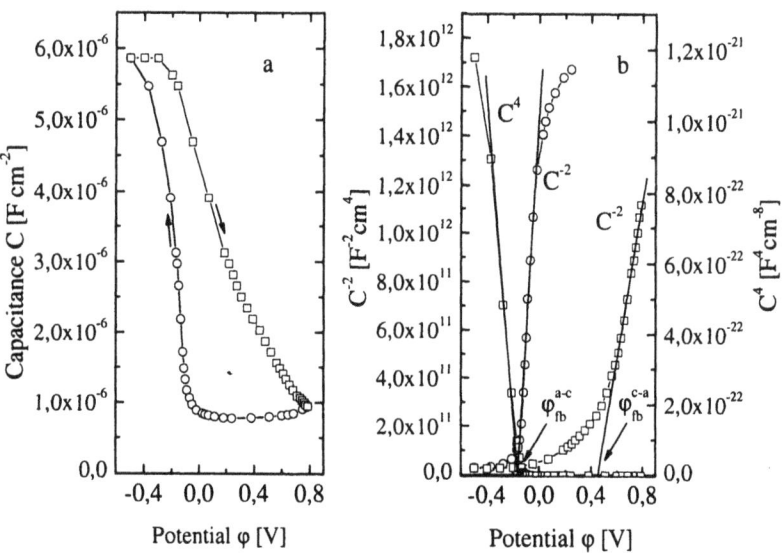

Figure 5.22 (a) $C_{exp}(\varphi)$ characteristic for TlBiSe$_2$ for surface 1, measured at "anode–cathode" ("a–c") and "cathode–anode" ("c–a") cyclic change of electrode potential; (b) same characteristic in reduced coordinates $C^4(\varphi)$ and $C^{-2}(\varphi)$.

fixed level of the capacitance in the cathode area of electrode potentials. The capacitance at the fixed level (C_i) is equal to $(5.0–7.2) \cdot 10^6$ F/cm^2. The hysteresis and the fixed level of the $C(\varphi)$ characteristics indicate oxidation of the surface of the sample, or the formation of a transition layer with dielectric (insulator) properties, which is characterized by the value

$$\frac{d_i}{\varepsilon_i} = \frac{\varepsilon_0}{C_i}, \tag{5.11}$$

where d_i is the thickness of the dielectric layer, ε_i is its dielectric permeability, and $\varepsilon_0 = 8.85 \cdot 10^{-14}$ F$\cdot$cm^{-1} is the dielectric constant. The experimental values of d_i/ε_i lie in the interval $(1.2–1.8) \cdot 10^{-8}$ cm.

The values of the flat-band potential (φ_{fb}) were defined from the $C(\varphi)$ characteristics in the depletion range of potentials which is linearized in $C^{-2}(\varphi)$ coordinates (Fig. 5.22b). The charge density (N_b) that built in a dielectric layer normalized to unity of the surface was also determined:

$$N_b = \frac{\Delta\varphi_{fb} C_i}{q}, \tag{5.12}$$

where $\Delta\varphi_{fb} = \varphi_{fb}^{a\text{-}c} - \varphi_{fb}^{c\text{-}a}$; $\varphi_{fb}^{a\text{-}c}$ and $\varphi_{fb}^{c\text{-}a}$ were measured at the anode-cathode $(a\text{-}c)$ and cathode-anode $(c\text{-}a)$ cycle change of the electrode potential; $q = 1.6 \cdot 10^{-19}$ C is the electron charge. The experimental values of N_b lie within the limits $(1.5–3.2) \cdot 10^{13}$ cm^{-2}. Using the flat-band potential values, it was possible to determine the surface potentials $V_s = -(\varphi - \varphi_{fb})$ (see Fig. 5.22b) and the ionized donor impurity concentration by the formula

$$N_d = \frac{2}{q\varepsilon_0\varepsilon_{sc}} \left(\frac{d(C_{sc}^{-2})}{dV_s} \right)^{-1}. \tag{5.13}$$

The experimentally obtained values of the concentration of ionized donor impurities N_d are listed in Table 5.1. For calculation we suggest $\varepsilon_{sc} = 21$ [126].

All experimental $C(V_s)$ characteristics at strong accumulation potentials are linear in $C^4(V_s)$ coordinates (see Fig. 5.22b). This indicates that the dispersion of the conduction band is close to parabolic, according to the formula (5.10).

TABLE 5.1

The Parameters for Surfaces 1 and 2 of TlBiSe$_2$

Parameter	Unit	Surface 1	Surface 2
φ_{st}^{before}	V	-0.100	$+0.190$
φ_{st}^{after}	V	$+0.250$	-0.185
N_d	cm^{-3}	$(0.7-2.3)\cdot10^{18}$	$(0.9-1.4)\cdot10^{18}$
$E_F - E_c$	eV	$0.00-0.030$	$-0.030--0.050$
m_e/m_0		$0.115-0.135$	$0.085-0.115$

The values of the Fermi level $E_F - E_c$ were determined (see Fig. 5.22b and Table 5.1) from the expression

$$E_F - E_c = -qV_s|_{C_{sc}=0} = \varphi|_{C_{sc}=0} - \varphi_{fb} \qquad (5.14)$$

The experimentally obtained values of $E_F - E_c$ are listed in the table.

The effective mass of electrons (m_e/m_0) was determined from the $\sim N^4(V_s)$ characteristics at strong accumulation potentials (see Fig. 5.22b and Table 5.1) by the formula

$$\frac{m_e}{m_0} = \frac{G^{-4/3}}{\varepsilon_{sc}^{2/3}} \left(\frac{dC_{sc}^4}{dv_s} \right)^{1/3}. \qquad (5.15)$$

Fair agreement of the experimental $C_{exp}(\varphi)$ characteristics with the theoretical $C_{sc}(V_s)$ characteristics is observed right up to the accumulation potential. The theoretical $C_{sc}(V_s)$ characteristics were calculated by self-consistent solution of the Schrödinger and Poisson equations. The experimentally obtained values of the effective mass of electrons and the ionized donor impurity concentration, $E_g = 0.45$ eV and $\varepsilon_{sc} = 21$ [125], the effective mass of holes in the valence band $m_h = 0, 40m_0$, and a spin-orbit splitting energy of the valence band of 1.0 eV were used in calculations.

The fixed level of the capacitance in the experimental $C_{exp}(\varphi)$ characteristics is determined by creation of a dielectric layer on the surface of TlBiSe$_2$.

There is no increase of capacity at the inversion range potential in the experimental $C_{exp}(\varphi)$ characteristics, in contrast to the theoretical one. This is explained by growth of the current through the TlBiSe$_2$-electrolyte interface. Physically, this means that the holes move into the electrolyte to maintain the electrochemical reaction at the surface of the semiconductor.

It is shown experimentally that the signs of φ_{st} for surfaces 1 and 2 differ before the field effect (see φ_{st}^{before} in Table 5.1). It is interesting that, following polarization in the field, the signs of φ_{st} are reversed (see φ_{st}^{after} in table). In our opinion, this change of sign indicates the predominance of one component of TlBiSe$_2$ which is different for surfaces 1 and 2. This result confirms the lamellar character of the original crystal, from which the 1 and 2 surfaces were obtained.

The results of experimental research and numerical simulations have allowed us to establish the following.

- On application of a cyclic perpendicular electric field to the TlBiSe$_2$ surface modulation of the differential capacitance of the system TlBiSe$_2$-electrolyte is observed.
- The experimental $C(\varphi)$ characteristics look like those typical of metal-insulator-semiconductor structures with an n-type semiconductor. It is possible to select an interval of potentials on the $C(\varphi)$ characteristics where they are linear in Schottky-Mott coordinates. This allows us to estimate the values of the flat-band potential and the donor impurity concentration (defects).
- Analysis of the experimental and theoretically calculated $C(\varphi)$ characteristics allowed us to estimate the value of the effective mass of electrons in the conduction band of TlBiSe$_2$.
- The hysteresis in the $C(\varphi)$ characteristics is caused by negative charge building up in the dielectric layer on the surface of TlBiSe$_2$.

5.2.6 (CdHg)Te

Surface properties of (CdHg)Te electrodes were studied in [28, 127–129]. The results of studies of capacitance-voltage and current-voltage characteristics in the FESE for $Cd_xHg_{1-x}Te$ ($x = 0.245$) are presented in Figure 5.23.

The capacitance of the electric double layer calculated on the basis of the current-voltage characteristic was found to be about 50 $\mu F/cm^2$. This is more than twice the value of the Helmholtz layer capacitance on the SE boundary and about an order of magnitude higher than the SCL capacitance, which, according to theoretical calculations carried out within the classical model of the SCL and taking account of degeneracy and nonparabolicity of the conduction band, should not exceed

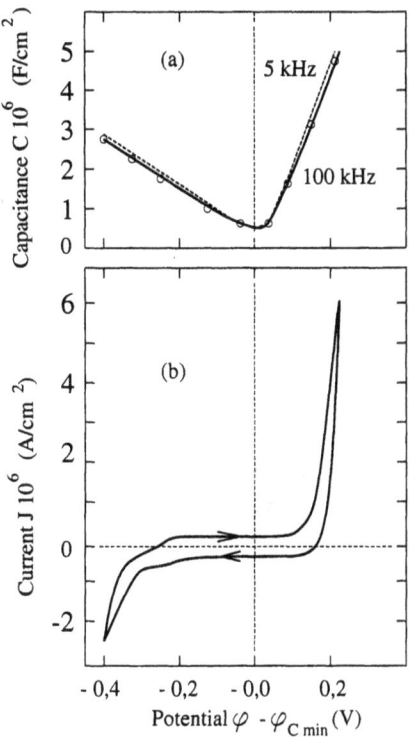

Figure 5.23 Capacitance-voltage (a) and current-voltage (b) characteristics of $Cd_xHg_{1-x}Te$ electrode ($x = 0.245$) in KCl aqueous solution in the range of ideal polarizability. Circles correspond to the capacitance-voltage characteristics calculated in the Kane model approximation for c band and parabolic model for v band.

$4-5\,\mu F/cm^2$. This discrepancy suggests that the system in question is actually a multilayer structure of the following type: (CdHg)Te–transition layer–electrolyte. The boundary (CdHg)Te–transition layer is, therefore, ideally polarized with a low density of electronic states on the (CdHg)Te surface. This is evidenced by the absence of any dependence of this boundary capacitance on the probing signal frequency. The electrophysical properties of the boundary are determined solely by the SCL of the (CdHg)Te electrode . At the same time, the boundary between the transition layer and the electrolyte represents a completely unpolarizable interphase boundary with an infinitely high capacitance.[1] The dependence of C_{min} on pH for two values

---

[1] The role of the transition layer in the system in question is likely to be played by Hg_2Cl_2 arising on the (CdHg)Te surface under the adsorption of chlorine from the electrolyte in the process of interfacial polarization. This assumption is confirmed by the coincidence of the ideal polarizability region of the specimen and that characterized by the electrochemical stability of Hg_2Cl_2 in a KCl solution. This assumption is also supported by the fact that substitution of the KCl solution by Na_2SO_4 and K_2SO_4

Figure 5.24 Minimum value of the capacitance of the $Cd_xHg_{1-x}Te$ electrode ($x = 0.245$) as a function of pH electrolyte for low- (solid line) and high-frequency (dashed line) measuring signals.

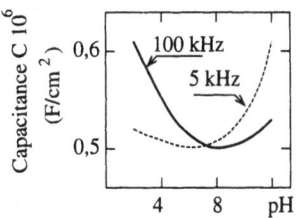

of frequency of the testing signal is presented in Figure 5.24. It is noteworthy that analogous dependence is observed for Ge, InAs, and InSb (see Figs. 6.4, 5.14, and 5.18).

Another means of realization of a (CdHg)Te–electrolyte boundary with an ideal polarizability is the employment of a nonproton electrolyte, namely, lithium perchlorite. This contains ortho-oxychinoline, producing complexes, which dissolve the products of electrochemical reactions on the (CdHg)Te electrode resulting from its polarization. This enables realization of an interphase boundary with low surface state density ($N_{ss} < 10^{11}$ cm^{-2}), hence extending the range of ideal polarizability of (CdHg)Te electrodes.

Studies of capacitance-voltage characteristics for narrow-gap $Cd_xHg_{1-x}Te$ for $0.195 < x < 0.4$ revealed that the electrode potential region where the conditions for ideal polarizability are satisfied incorporates potentials causing degeneracy of electrons and holes on the surface [128]. This allows us to find such band structure parameters as the energy distribution of differential densities of electron (hole) states, dN/dE, and their effective masses in the c and v bands in the near-surface regions of these compounds. Figure 5.25 shows the experimental electron state densities as a function of energy for two compositions of (CdHg)Te. Here, the relevant theoretical curves calculated within nonparabolic (Kane) and a parabolic dispersion approximations are given. The experimental curves coincide, within the limits of measurement error, with the theoretical ones calculated within a nonparabolic dispersion approximation, for $Cd_xHg_{1-x}Te$ with x in the range $0.195 < x < 0.260$. Thus, for the compounds considered, the c band in the SCL region, as well as in the bulk, is described

---

resulted in an increase of the surface state density and, as a result, in a narrowing of the ideal polarizability region of the (CdHg)Te-electrolyte system. Within this model, the system Hg_2Cl_2–concentrated KCl solution proves to be an analogue to the standard calomel electrode (see, for example, [3, 130]).

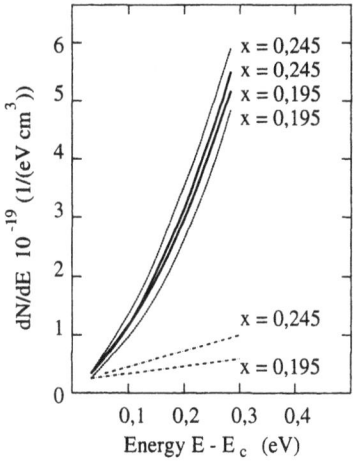

Figure 5.25 Energy distribution of the differential density of electronic states in the c band of $Cd_xHg_{1-x}Te$: solid line, experimental dependences; dashed line, calculated with Kane dispersion law approximation; dot-dashed line, calculated within parabolic approximation.

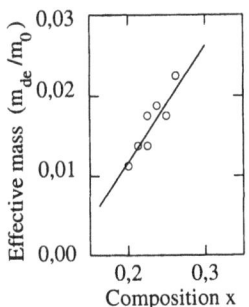

Figure 5.26 Variation in the effective mass of the electronic state density in $Cd_xHg_{1-x}Te$ electrode with x. Solid line is the theoretical calculation; circles are experimental data.

by a nonparabolic (Kane) dispersion law. The effective masses of the states density in the conduction band in the SCL, calculated from the $C(\varphi)$ curves in the region of V_s corresponding to strong electron gas degeneracy, were found to be in good agreement, within the experimental error, with the bulk values (Fig. 5.26).

The values of $E_F - E_i$ for compositions within $0.195 < x < 0.26$ obtained from the $C(\varphi)$ characteristics and their dependence on composition, calculated from the neutrality equation within the Kane model for the conduction band and within the parabolic one for the valence band of heavy holes, are shown in Figure 5.27. In calculations, the effective mass of the heavy hole for all investigated (CdHg)Te compounds was taken to be $0.45m_0$ [88, 131]. The agreement, accurate to k_0T/q, between the experimental values of $E_F - E_i$ and the theoretical ones suggests that the v band in the SCL for $Cd_xHg_{1-x}Te$

Figure 5.27 Variation in Fermi level location in $Cd_xHg_{1-x}Te$ electrode with x. Solid line is the theoretical calculation; circles are experimental data.

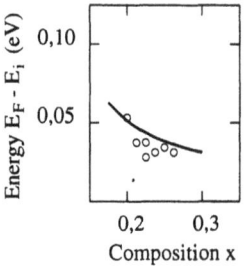

Figure 5.28 Minimum capacitance as a function of temperature for $Cd_xHg_{1-x}Te$ ($x = 0.21$).

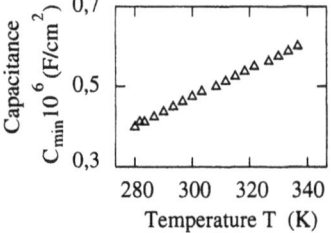

within the whole interval of x ($0.195 < x < 0.26$) can be described by the parabolic dispersion law with the value of the heavy-hole effective mass close to the bulk one ($m_{hh} = 0.45m_0$). The c band can be characterized by the Kane model. The linear temperature dependence of C_{min} for $x = 0.21$ also points to the nonparabolic character of the c band in such materials. Its slope yields the matrix element of the quasi-impulse P which takes into account the interaction between the c and v bands in the SCL of (CdHg)Te (Fig. 5.28). This value was found to be $4.1 \cdot 10^{-8}$ eV·cm, which is about two times lower than the bulk value for this material, known from literature [131].

Data on semiconductor parameters such as E_g and m_{de} can also be obtained from measurements of the intrinsic concentration (n_i) of charge carriers in a semiconductor. Figure 5.29 shows the values of intrinsic concentration of electrons in $Cd_xHg_{1-x}Te$ at different values of x in the range (0.195–0.4) obtained from the capacitance values at the minimum of the $C(\varphi)$ characteristic. For comparison, we give the data obtained for bulk $Cd_xHg_{1-x}Te$ at 300 K [128]. They demonstrate that, for compounds with x varying between 0.195 and 0.26, the experimental values of n_i defined from $C(\varphi)$ dependences and characterizing the near-surface SCL region are in good agreement with the values of n_i in the bulk crystals. For $Cd_xHg_{1-x}Te$ with x from 0.26 to 0.40, the experimental values of n_i derived from the

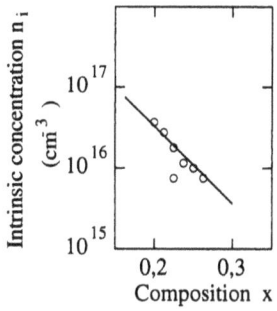

Figure 5.29 Variation in the intrinsic concentration of electrons with x. Solid line is the theoretical calculation for the bulk; circles are experimental data.

capacitance measurements were found to be several times lower than those obtained for the bulk semiconductor. Such a discrepancy between the values of n_i obtained from the $C(\varphi)$ characteristics for the near-surface region and those characterizing the bulk cannot be accounted for either by the presence of surface states or by a surface oxide layer which could affect the value of C_{min} and hence bring about an error in the determination of n_i. The authors of [128] suggest that this lack of agreement allows a conclusion that (CdHg)Te compounds vary in composition on going from the near-surface region to the interior. It is assumed, herewith, that the near-surface region is characterized by a higher forbidden band value as compared with the bulk one. This is in accordance with the works [132–135] where the possibility was demonstrated of exhaustion of the near-surface volume of (CdHg)Te by mercury, resulting in increase of the forbidden band width.

5.2.7 (ZnHg)Te

The fundamental investigations of the electrophysical properties of (ZnHg)Te electrodes were carried out on epitaxial films of different compositions [136]. FESE measurements showed that the ideal polarizability region lies within the potential range of $-0.350 < \varphi - \varphi_{min} < 0.250$ V, where φ_{min} is the electrode potential related to the minimal capacitance value in the $C(\varphi)$ characteristics. The current-voltage curves (Fig. 5.30a) are similar in character to "charging" curves [3]. The capacitance-voltage curves (Fig. 5.30b) are independent of the direction and the rate of the electrode potential change within the range of 10–100 mV/s and the capacitance does not feature frequency dispersion in the range of a probing signal of 5–100 kHz. This is indicative of the fact that (ZnHg)Te electrode surfaces are free of

Figure 5.30 Current-voltage (a) and capacitance-voltage (b) curves for the system i-$Zn_{0.17}Hg_{0.83}Te$–1.5 mol KCl aqueous solution with different frequencies of the probing signal: (1), 5 kHz; (2), 100 kHz.

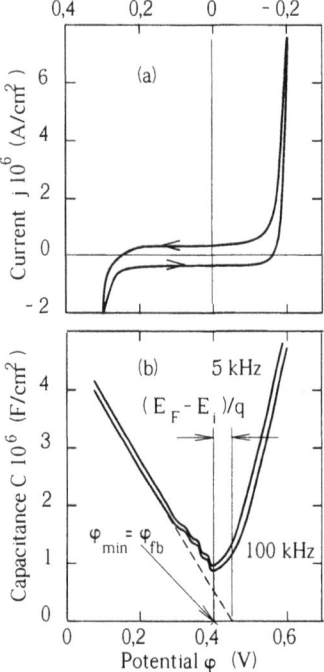

surface states, capable of recharging during the range of relaxation time $10^{-5} < \tau < 10^{-3}$ s and the capacitance of the (ZnHg)Te-electrolyte boundary is completely defined by the electrode SCL capacitance. Capacitance-voltage characteristics of $Zn_xHg_{1-x}Te$ films show a region that is straightening in the Schottky-Mott coordinates, which enables a determination of φ_{fb} as well as of the type and concentration of ionized impurities. This region is most extended for a composition of $x = 0.20$–0.22, corresponding to the maximum value of E_g. The left-side branches of the capacitance-voltage characteristics at $\varphi > 0.2$ for all (ZnHg)Te films investigated correspond to a strong degeneracy of the electron gas in the SCL of the semiconductor (see Fig. 5.31). The Fermi level position at the electrode surface can be obtained by making use of the relation $qV_s = E_F - E_i$ which is valid for $C_{sc} = 0$. For compositions with $x = 0.15$, 0.16, 0.17 the values of $E_F - E_i$ were equal to 0.050 ± 0.010 eV, which corresponds to the calculated values for an intrinsic material at room temperature; for $x = 0.20$–0.22, $E_F - E_i = -0.100$ eV. This is in accordance with the Fermi level position in the bulk of a material with the acceptor

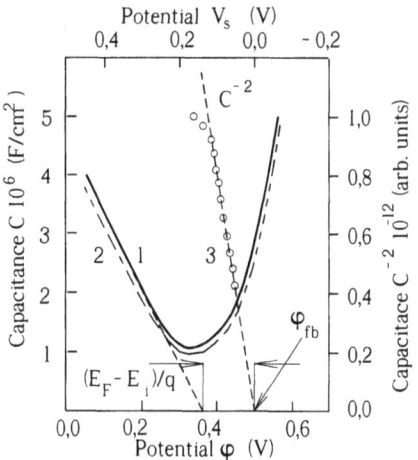

Figure 5.31 Capacitance-voltage characteristics of the $Zn_xHg_{1-x}Te$ ($x = 0.20$–0.22, p type, $N_a = 1,5 \cdot 10^{18}$ cm^{-3}) electrode in 1.5 mol KCl aqueous solution: (1), experimental curve; (2), calculated curve; (3), experimental curve in Schottky-Mott coordinates.

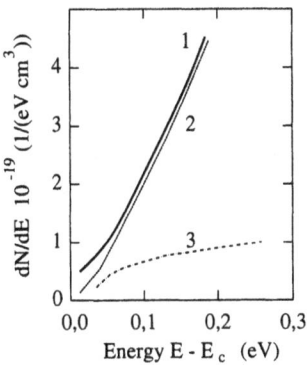

Figure 5.32 Differential density of electron states in c band as a function of electron energy for $Zn_{0,17}Hg_{0,83}Te$ electrode: (1), experimental curve; (2), theoretical curve in Kane approximation; (3), theoretical curve in parabolic approximation.

density of $N_a = 1.5 \cdot 10^{18}$ cm^{-3} calculated from the electroneutrality equation.

The employment of the FESE technique first provided an opportunity to find the characteristics of the electron band structure of $Zn_xHg_{1-x}Te$ for a variety of values of x. With reference to the latter, we note that a linear dependence of the capacitance on the potential in the electron gas degeneracy region points to the Kane dispersion law in the c band. The same law follows from the juxtaposition of experimental curves $dN/dE \sim (E - E_c)$ and the theoretical ones calculated within the Kane and parabolic approximations (Fig. 5.32). Also note that the assessment of the effective mass of the state density in the c band putting $\varepsilon_{sc} = 17$–20 [137] yielded a value close to the bulk one.

Summarizing the results of investigations using FESE [136] it may be concluded that the composition of narrow-gap (ZnHg)Te epitaxial films remains the same through the specimen up to the surface, with the conduction band being nonparabolic (Kane type) and the valence band for heavy holes being parabolic.

5.2.8 (MnHg)Te

Measurements of capacitance-voltage characteristics of $Mn_xHg_{1-x}Te$ electrodes for $x = 0.15$ (Fig. 5.33) showed that, for this semiconductor, the c band in the SCL region is characterized by a nonparabolic (Kane-type) dispersion law [138]; the electron effective mass was found to be $m_e = 0.055m_0$, in good agreement with the bulk data for this composition. The valence band is characterized by a parabolic dispersion law with $m_h = 0.55m_0$. The effect of a transverse magnetic field ($B = 2$ T) on the $C(\varphi)$ characteristics was revealed, which may be of interest because of the high magnetosensitivity of these compounds, and, hence, the potentialities for their employment for microelectronics applications [88]. The most pronounced capacitance change in the magnetic field was observed at the minimum of the $C(\varphi)$ curve. The capacitance was observed to decrease in magnitude; the decrease was about 2–3% with the accuracy of measurement being 1%, which can be assigned to the cyclotron resonance broadening of E_g. Indeed, in a magnetic field applied normal to the electrode surface, the electron energy can be expressed as $E = \hbar\omega_{e(h)}(n + 1/2) + \hbar^2 k^2/2m_{e(h)}$. Note that, formally, for carriers moving along the surface, it is necessary to consider $E_g + \hbar(\omega_e + \omega_h)/2$, rather than E_g. Then, for flat bands,

Figure 5.33 Capacitance-voltage characteristics of $Mn_xHg_{1-x}Te$ electrode ($x = 0.15$) in 1 mol KCl aqueous solution: (1), without magnetic field; (2), in transverse magnetic field $B = 2$ T.

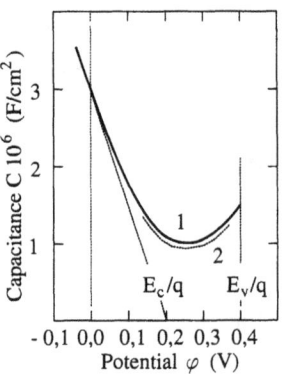

that is, at the minimum of the $C(\varphi)$ curve, the expression for the charge density $\rho_B(\varphi) \sim \exp[-E_g/(2k_0 T)]$ must be substituted by the following one:

$$\rho_B(\varphi) \sim \exp\left(-\frac{\hbar(\omega_e + \omega_h) + 2E_g}{4k_0 T}\right), \qquad (5.16)$$

where $\omega_e = qB/m_e$ and $\omega_h = qB/m_h$. Making use of the general expression for the differential capacitance of the semiconductor SCL in the form

$$C_{sc}(\varphi) = \pm\sqrt{\frac{\varepsilon_0 \varepsilon_{sc}}{2}} \rho(\varphi) \left(-\int_0^\varphi \rho(\varphi)\, d\varphi\right)^{-1/2}, \qquad (5.17)$$

we arrive at the following expression:

$$\Delta C/C_{min} \simeq 1 - \exp\left(-\frac{\hbar(\omega_e + \omega_h)}{4k_0 T}\right). \qquad (5.18)$$

Assuming $m_e = 0.055m_0$, $m_h = 0.35m_0$, and $B = 2\,$T, we have $\Delta C/C_{min} \simeq 2\%$, which is in fair agreement with experiment.

The results of the investigations for other compositions $(0 \leqslant x \leqslant 0.24)$ of $Mn_x Hg_{1-x}Te$ electrodes are given in [129].

5.2.9 (PbSn)Te

Representative capacitance- and current-voltage characteristics for (PbSn)Te compounds for the ideal polarizability region are given in Figure 5.34 [120].

An estimate of the maximum value of the total capacitance of the (PbSn)Te-electrolyte boundary from current-voltage characteristics gives a value about $20\,\mu F/cm^2$, which is close to that of the Helmholtz layer. On the $C(\varphi)$ curve, one can distinguish a capacitance region that straightens in the Schottky-Mott coordinates $(C \sim \varphi^{1/2})$. This made possible a determination of the flat-band potential which was found to be $\varphi_{fb} = -0.94\,$V, and also the acceptor density in the electrode material $(N_a = 2.5 \cdot 10^{18}\,cm^{-3})$. The latter value is in agreement with the specifications of the given sample and the values of N_a measured using the Hall method. Juxtaposition of φ_{fb} and φ_0 for the unpolarized state of the electrode shows that in the absence of polarization the surface of the p-(PbSn)Te electrode is rich in holes.

69

Figure 5.34 Capacitance-voltage (a) and current-voltage (b) characteristics of (PbSn)Te electrode in KOH + EDTA aqueous solution.

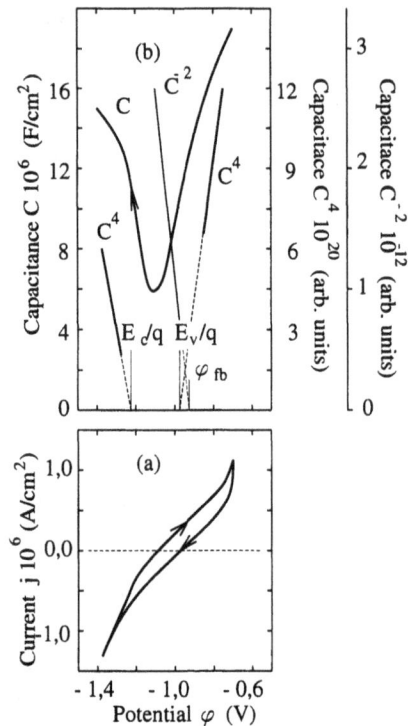

At large values of φ, corresponding to the accumulation and the inversion of charge carriers on the (PbSn)Te surface, the $C(\varphi)$ curves are well straightened in the $C^4 \sim \varphi$ coordinates. This is indicative of the electron and hole degeneracy in the SCL region within this interval of φ and of the parabolic character of the dispersion law for the allowed bands in the SCL region of (PbSn)Te. This is also evidenced by the estimate of the value of E_g obtained from extrapolation of the $C(\varphi)$ dependence up to the intersection with the φ axis, yielding $E_g = 0.21$ eV, in good agreement with the data known from the literature for bulk $Pb_{1-x}Sn_xTe$ with $x = 0.2$ at $T = 300$ K [88]. The effective masses of the electron and hole state densities in the relevant allowed bands near the (PbSn)Te surface are almost the same and close to their bulk values [88].

5.3 ZERO-GAP SEMICONDUCTORS AND SEMIMETALS

Experimentally, a gapless (zero-gap) state was first observed in HgTe crystals in the mid-1950s by Tsidil'kovsky [88]. Since then, the materials featuring this property have been studied extensively (see, for example, [131, 139, 140]). For practical reasons, a major feature of zero-gap semiconductors is the extreme sensitivity of their electron systems, in particular, the electron and hole concentrations, to external perturbations: temperature, pressure, electric and magnetic fields, and light illumination. Zero-gap semiconductors may be used as active elements of waveguides in the submillimeter and far-infrared regions, controlled by magnetic and electric fields and by temperature. Uses include emission amplitude modulators, phase rotators, switching devices, attenuators, and so on [88].

Basically, a characteristic feature of these materials is the fact that their band structure is mainly defined by relativistic effects whose role is enhanced on increasing the atomic number. As a result, these compounds provide the basis for studies of their effect on the formation of the band structure of solids.

5.3.1   HgTe and $Cd_xHg_{1-x}Te$ $(x = 0.03-0.05)$

In a series of works (see [129, 141–143]), it was demonstrated that a combination of a previous treatment of a HgTe surface (chemicodynamical polishing with subsequent chemical etching in a solution of Br in methanol) and a proper choice of the electrolyte provides a wide range of electrode potentials $[-0.30 < (\varphi - \varphi_{fb}) < +0.30\,V]$, in which the conditions of ideal polarizability are fulfilled with capacitance-voltage and current-voltage curves being stable and well reproducible.

Typical dependences of capacitance and current densities on the electrode potential for the HgTe–aqueous KCl solution boundary are shown in Figure 5.35. As seen from the figure, the $C(\varphi)$ curves feature neither frequency dispersion nor hysteresis. The current-voltage characteristic is typical for an ideally polarized electrode. The analysis of these results indicates that the capacitance minimum in the $C(\varphi)$ curve corresponds, within the notion of the SCL for HgTe and other zero-gap semiconductors, to the flat-band state on the surface. An increase of the surface potential results in the accumulation of electrons in the completely degenerate conduction band, while a decrease of the former brings about the accumulation of holes. The minimal capacitance value for an intrinsic semiconductor (HgTe is intrinsic at $T = 295\,K$) permits the assessment of the intrinsic density of charge carriers (n_i). The latter was found to be $2.7 \cdot 10^{17}\,cm^{-3}$ for HgTe, which is in close agreement with the known datum for the bulk

Figure 5.35 Differential capacitance (a) and polarization current (b) of the HgTe electrode as functions of the electrode (surface) potential in 2 mol KCl solution: (1) and (1′) measurements at 100 kHz and 5 kHz, respectively; (2), calculations carried out using the model of an ideal zero-gap semiconductor (three-band model approximation); (3) calculations carried out using the model of a zero-gap semiconductor with a fluctuation potential.

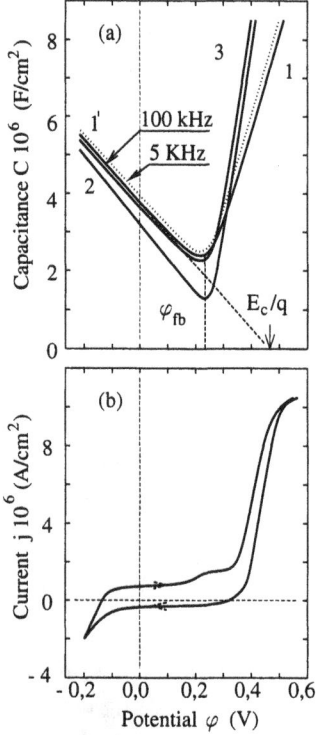

$(2.0 \cdot 10^{17} \, \text{cm}^{-3})$ [88, 131]. The position of the capacitance minimum in the $C(\varphi)$ curve on the electrode potential scale, corresponding to $q \cdot \varphi_{\text{fb}} = (E_F - E_i)$, with respect to $V_s = 0$ yielded a value of 0.1–0.32 V. Using the relation between the electrochemical scale and the physical one for surface potentials and with knowledge of the $E_c - E_F$ value, the electron affinity for HgTe can be obtained. The latter was found to be 4.7 eV, which is in good agreement with the known datum 4.3–4.7 eV [131].

As the anode polarization of the HgTe electrode is increased, a number of U-shaped regions of reversible change of the $C(\varphi)$ curves occur, whose minimum is shifted to the region of positive values of the electrode potential, with the value at the minimum being unchanged $(1.1 \, \mu\text{F/cm}^{-2})$ (Fig. 5.36). The same is true of the value of C_0 corresponding to the unpolarized state, which was changed in the same direction. The capacitance-voltage characteristic taken for $\varphi = 1.3 \, \text{V}$ shows a region decaying away into the anodic polarization region

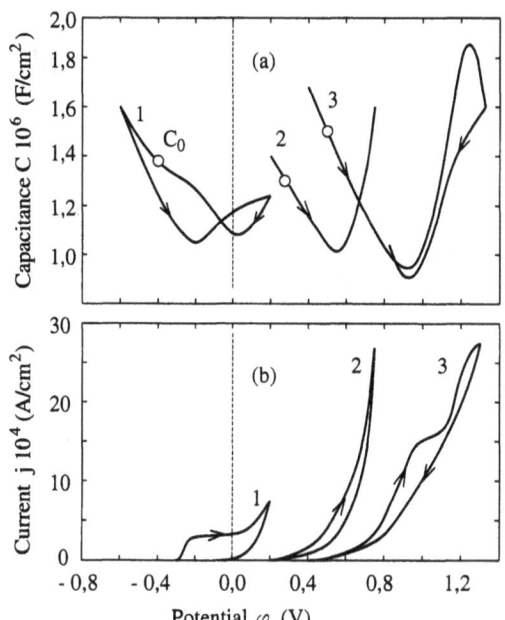

Figure 5.36 Capacitance-voltage (a) and current-voltage (b) characteristics of HgTe electrode in aprotonic electrolyte.

which, according to [141], is typical for all semiconductors at the stage of formation of anode oxide on their surface.

The results of FESE studies for $Cd_xHg_{1-x}Te$ zero-gap semiconductors ($x = (0.03–0.05)$) are shown in Figure 5.37. An estimate of the intrinsic density of charge carriers on the basis of the $C(\varphi)$ characteristics yielded the value of $1.8 \cdot 10^{17}$ cm^{-3}, which is close to the bulk value of $1.3 \cdot 10^{17}$ cm^{-3} [143].

The electron branches of the experimental $C(\varphi)$ curves for $Cd_xHg_{1-x}Te$ ($x = 0.03–0.05$) and HgTe were used for the determination of the electron parameters of the SCL. The linearity of the electronic branches is an indication of the nonparabolic character of the conduction band dispersion law in the SCL region of these materials. That made it possible to define the effective masses of the states density in the c band in the region under consideration. They were found equal to $0.004m_0$ and $0.011m_0$, respectively. These values are close to the bulk ones for these compounds; see [88, 131].

For zero-gap semiconductors as well as for those where the gap is close to zero, the dispersion laws for the c and v bands of light holes become ultrarelativistic and are described by the expression $E = sp$,

Figure 5.37 Differential capacitance
(a) and polarization current (b) of the
$Cd_xHg_{1-x}Te$ ($x = 0.03$–0.05)
electrode as functions of the
electrode (surface) potential in 2 mol
KCl solution: (1) and (1′)
measurements at 100 and 5 kHz,
respectively; (2), calculations carried
out using the model of an ideal
zero-gap semiconductor (three-band
model approximation); (3),
calculations carried out using the
model of a zero-gap semiconductor
with a fluctuation potential.

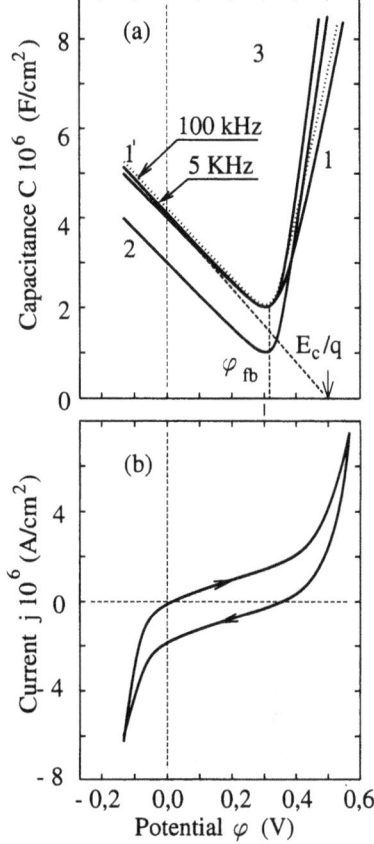

where $s = E_g/2m_{de}$ [88].[2] It can be shown [143] that, in the ultra-relativistic approximation for the dispersion of the conduction band in the case of strong degeneracy of carriers, the differential capacitance should be linear with the surface potential as in the Kane-type dispersion case:

$$C_{sc} = 4q^2 \left(\frac{\pi \varepsilon_0 \varepsilon_{sc}}{3\hbar^3} \right)^{1/3} \left| \frac{E_g}{2m_{de}^*} \right| V_s. \tag{5.19}$$

If one knows E_g and ε_{sc}, from the experimental $C(\varphi)$ dependences one may define with this formula the effective masses of the states

---

[2] In the case of zero-gap semiconductors, by E_g one means the energy spacing between the conduction band and the valence band for light holes at $k = 0$ [88].

density in the conduction band. The estimates made showed that, for $Cd_xHg_{1-x}Te$ $(x = 0.03–0.05)$ and HgTe, the effective masses of the states density in the SCL for the c band, obtained from the experimental $C(\varphi)$ curves, assuming an ultrarelativistic dispersion law, are close to those calculated (within the Kane approximation) for the bulk values [88]. That agreement shows that, for these materials, the dispersion law of the c band in the SCL is the same as in the bulk and close to the ultrarelativistic one.

From the electron branches of the experimental $C(\varphi)$ curves, the Fermi level positions $(E_F - E_i)$ for the investigated HgTe (Fig. 5.35) and $Cd_xHg_{1-x}Te$ $(x = 0.03–0.05)$ (Fig. 5.37) were determined; they were found to be 0.230 and 0.150 eV, respectively.[3] These values, for both HgTe and $Cd_xHg_{1-x}Te$ $(x = 0.03–0.05)$, were higher than those calculated within the three-band model, by 0.110 and 0.70 eV, respectively.[4]

Analysis of the temperature dependence of $C(\varphi)$ curves in the interval $213 < T < 320 \, K$ showed [143] that, for $Cd_xHg_{1-x}Te$ $(x = 0.03–0.05)$ and HgTe, the minimal capacitance is a linear function of temperature (see Figs. 5.38a and 5.38b, respectively). This points to the fact that both bands on the surface of zero-gap semiconductors are extremely nonparabolic and also supports the idea of the nonparabolicity of the valence band in zero-gap semiconductors put forward in [143]. The dependence of C_{min} on temperature allows one to estimate the value of the matrix element of the quasi-impulse. For both types of materials it was found to be $4.0 \cdot 10^{-8} \, eV \cdot cm$, two times less than the bulk value [88].

5.3.2 HgSe

We note that compounds based on mercury telluride have been studied in sufficient detail; compounds based on mercury selenides call for further investigation. The fact that there are only a few papers dedicated to HgSe surface properties is explained by methodological difficulties in forming stable insulating and passivating coatings at the material surface [144–146]. This, in turn, does not allow the efficient application of sophisticated conventional methods based on

---

[3] Here, by E_i the mid-spacing between the c band and v band of light holes is meant.

[4] In the calculations, we used the following values: $E_g = 0.050 \, eV$, $m_{de} = 0.004 m_0$ for $Cd_xHg_{1-x}Te$ $(x = 0.03–0.05)$ and $E_g = 0.110 \, eV$, $m_{de} = 0.010 m_0$, $m_{hh} = 0.45 m_0$, $m_{lh} = 1.5 m_0$, and $\varepsilon_{sc} = 21.0$ for HgTe [88, 131].

Figure 5.38 Temperature dependences of the capacitance at different values of electrode potential for $Cd_xHg_{1-x}Te$ ($x = 0.03–0.05$) (a) and HgTe electrodes (b).

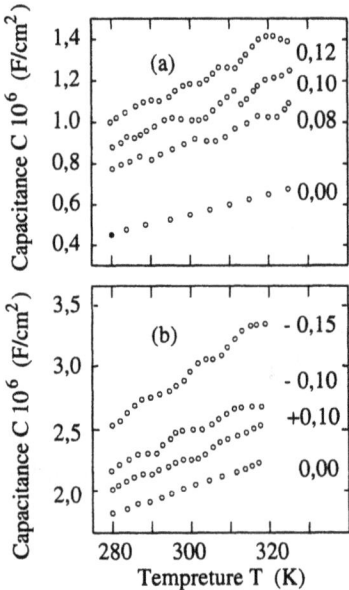

the analysis of metal-insulator-semiconductor structures or Schottky barriers to study the electrical properties of the material surface. Moreover, the papers dedicated to the bulk HgSe band parameters (as a rule, such data involve low-temperature measurements) do not offer a unified concept of these parameters (we recall the recent discussion about the HgSe band structure [145–148]). For example, the effective electron mass is $m_e = 0.05m_0$ at $T = 300$ K at the Kane band gap $E_g = -0.2$ eV [148]; $m_e = (0.015–0.019)m_0$ and $E_g = -0.2$ eV; $m_e = 0.019m_0$ at $T = 95$ K and $E_g = -0.22$ eV at $T = 300$ K; and $m_e = 0.019m_0$ and $E_g = -0.22$ eV at $T = 300$ K.

According to [147], the parameters for the $Hg_xSe_{(1-x)}$ binary compounds at $T = 300$ K are $E_g = -0.18$ eV, $m_e = 0.001m_0$ ($x = 0.05$) and $E_g = -0.2$ eV, $m_e = 0.008m_0$ ($x = 0.1$). According to [148], the matrix element of the quasimomentum operator for HgSe is $P = 7.5 \cdot 10^{-8}$ eV·cm.

We measured and analyzed the electrical properties of the HgSe surface and subsurface regions (the bulk density of electrons in the studied sample was $n = 4.1 \cdot 10^{17}$ cm^{-3} determined from the Shubnikov–de Haas oscillation) using the field-effect method in electrolytes [149].

The electrical and band parameters of the HgSe space charge layer were determined by capacitance-voltage [$C(\varphi)$] characteristics in a system of HgSe and a saturated aqueous solution of KCl. Simultaneously with the $C(\varphi)$ characteristics, current-voltage $[I(\varphi)]$ characteristics were measured to control currents through the semiconductor-electrolyte interface. All the measurements were carried out in a constant-temperature (at $T = 295\,\text{K}$) electrochemical cell. The HgSe sample surface was polished with diamond paste and then chemicodynamically polished in an 8% bromine solution in methanol. Immediately before measurements, the surface was etched in a (2–8)% bromine solution in methanol. The differential capacitance was measured using a rectangular pulsed signal with the test pulse duration $\tau = 1\,\mu\text{s}$ with a cyclic sweep of the electrode potential V at the rate of 10–30 mV/s. The electrode potential was measured with reference to a carbon-graphite electrode.

All the $C(\varphi)$ measurements were carried out in an electrode potential range with the following properties.

- The current through the interface was negligible; that is, there was no significant contribution of the current component caused by electrochemical reactions proceeding when the semiconductor-electrolyte interface was polarized due to the field effect.
- The $C(\varphi)$ characteristics remained unchanged during the multiple cyclic variation of the electrode potential (for a few hours).
- The HgSe-electrolyte interface had no surface states recharged within relaxation times $\tau \geqslant 10^{-5}$ s.
- No capacitance saturation was observed, which would indicate formation of a surface oxide with a thickness comparable to the Debey screening radius in HgSe. If such a layer exists at the semiconductor surface, its thickness is comparable to the Helmholtz layer thickness (i.e., thinner than 0.2–0.4 nm) and certainly thinner than the SCL width. If these conditions are met, the total change of the electrode potential when sweeping the voltage over the cell falls within the HgSe SCL ($|\Delta\varphi| = |\Delta V_s|$, where V_s is the surface potential), and the measured capacitance is the differential capacitance of the semiconductor SCL.

One can see from Figure 5.39 that the experimental $C(\varphi)$ characteristic is linearized in the range of sufficiently high negative electrode potentials (the electron portion of the characteristic). It can be readily shown that exactly such a dependence of the specific capacitance of the semiconductor SCL should be observed for Kane semiconductors in the region of band bendings exceeding the Kane band gap E_g [149].

77

Figure 5.39 Experimental
capacitance-voltage characteristic
of the system consisting of HgSe
and saturated solution of KCL.

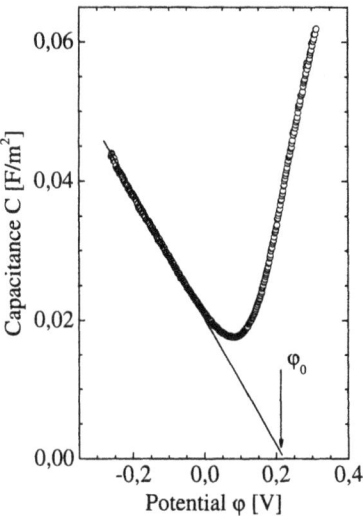

The slope of the $C(\varphi)$ characteristic is directly defined by the matrix element P of the quasi-momentum operator. Indeed, for pronounced band bendings, the dispersion law in the two-band approximation

$$E = \sqrt{(2P^2 k^2 / 3 + E_g^2 / 4)} - E_g / 2$$

is resonably adequate for narrow- and zero-gap semiconductors and is close to linear for most of the enriched-layer electrons involved in screening, $E \approx \sqrt{2/3} Pk - |E_g/2|$. Hence, the density of states

$$D(E) = \frac{2}{(2\pi)^3} \int\limits_S \frac{dS}{\nabla_k E}$$

in the conduction band takes the form

$$D(E) = (3/2)^{3/2} \frac{(E + |E_g/2|)^2}{\pi^2 P^3}.$$

The last expression includes only the first-order terms in the series expansion in the parameter $E_g/2E$; dS is the isoenergetic surface element in the quasi-wave-vector space. Since the conditions of pronounced degeneracy $\mu_s \geq \mu \geq \mu_b \gg kT$ ($\mu_s = qV_s + \mu_b$, μ and μ_b are, respectively, the Fermi levels at the surface, in the SCL, and in the

78

bulk of the semiconductor) are met in almost the entire enriched layer of the considered material, the local carrier concentration is given by

$$n(\mu) = \int_0^\mu D(E)dE = (3/8)^{1/2} \frac{(\mu + |E_g/2|)^3}{\pi^2 P^3}.$$

Within this approximation, which is adequate in a wide range of experimental conditions, the SCL differential capacitance is written as

$$C = \frac{dQ_s}{dV_s},$$

where the surface charge density in the SCL

$$Q_s = \varepsilon_0 \varepsilon_{sc} \left. \frac{dV}{dz} \right|_{V=V_s} = \left. \frac{\varepsilon_0 \varepsilon_{sc}}{q} \frac{d\mu}{dz} \right|_{\mu=\mu_s}$$

(ε_{sc} is the dielectric constant of the semiconductor), is defined by the first integral of the Poisson equation,

$$\left. \frac{d\mu}{dz} \right|_{\mu=\mu_s} = \sqrt{\frac{2q^2}{\varepsilon_0 \varepsilon_{sc}}} \left(\int_{\mu_b}^{\mu_s} n(\mu)d\mu \right)^{1/2}.$$

Elementary calculations then yield the following formula for C:

$$C = \beta \frac{(qV_s + \mu_b + |E_g/2|)^3}{\sqrt{(qV_s + \mu_b + |E_g/2|)^4 - (\mu_b + |E_g/2|)^4}}.$$

Here,

$$\beta = \sqrt{\frac{q^2 \varepsilon_0 \varepsilon_{sc}}{(2/3)^{1/2} \pi^2 P^3}}.$$

Expanding the above expression for the capacitance in series in the small parameter $(\mu_b + |E_g/2|)/qV_s$, we arrive at the simple expression

$$C \approx \beta \left\{ qV_s + \mu_b + |E_g/2|) \left[1 + \frac{1}{2} \left(\frac{\mu_b + |E_g/2|}{qV_s} \right)^3 \right] \right\}.$$

Since the correction related to deviation from the quasiultrarelativistic limit $E_g = 0$, $\mu_b = 0$ arises only in the third order with respect to $(\mu_b + |E_g/2|)/qV_s$, the contribution of the term in square brackets is insignificant even at relatively small band bendings and decreases rapidly as V_s increases, and we arrive at the sought-for linear dependence

$$C \approx \beta(qV_s + \mu_b + |E_g/2|).$$

The derivative dC/dV_s, measured in the linear electron portion of the $C(\varphi)$ characteristic, is defined by only two material parameters, that is, the matrix element P of the quasi-momentum operator and the dielectric constant. This is ultimately a direct consequence of the ultrarelativistic character of electron motion in the enriched layer, whose dispersion $E \approx \sqrt{2/3}(P \cdot \mathbf{k})$ is described by the single parameter P. Using the experimental value $dC/d\varphi = dC/dV_s$ for the electron portion of the $C(\varphi)$ characteristic (see Fig. 5.39) and the dielectric constant $\varepsilon_{sc} = 25.6$ from [148], we find that the matrix element of the quasi-momentum operator is $P = 8.2 \cdot 10^{-8}$ eV·cm for HgSe . This value conforms well to the data of [147] for HgSe at low temperatures and is close to P in HgTe. On the assumption that the zero electrode potential corresponds to the condition of flat bands, $\varphi_{fb} = V_s = 0$, the cutoff potential $\varphi|_{C=0} = 0.220$ Å, found by extrapolation of the linear portion of the $C(\varphi)$ characteristic to $C = 0$, can be used to estimate $\mu_b + |E_g/2|$. If we use the Kane band gap $E_g = -0.220$ eV [147, 148], the Fermi level in the bulk is $\mu_b \approx 0.110$ eV, which conforms well to the value $\mu_b = 0.095$ eV, determined from the electroneutrality equation with the above values of P, E_g, and n.

5.3.3 Carbon

Studies of surface properties of these materials, treated as electrodes, were made in the pioneering work of Gerischer [150]. In this work it was first shown that the differential capacitance of the electrolyte–graphite boundary is defined by the SCL capacitance of graphite and reflects the character of electronic processes on its surface.

The results of FESE studies, obtained using a high-temperature annealing of polymer materials (Vitreous Carbon type SU-2500) and pyrographites, are represented in [151, 152]. A KCl solution in ethylene glycol was used as the electolyte. Measurements showed that the current-voltage curves in the range of $-0.8 < \varphi < 0.8$ V resemble

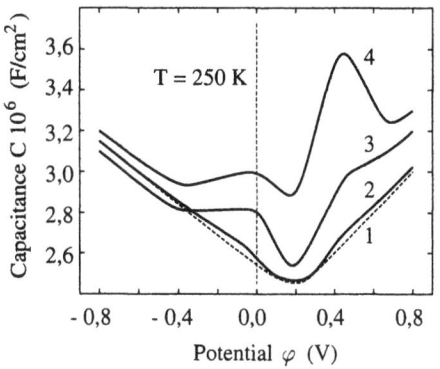

Figure 5.40 Capacitance-voltage characteristics of the vitreous carbon electrode in 1 mol KCl aqueous solution.

the curves corresponding to the charging of the geometrical capacitance calculated for the double electric layer $C = j \, (\mathrm{d}t/\mathrm{d}\varphi) = 1.75 \cdot 10^{-5} \, \mathrm{F/cm^2}$. The capacitance-voltage curves were U shaped with $C_{min} = 3.0 \cdot 10^{-6} \, \mathrm{F/cm^2}$ for specimens made of pyropraphite and with $C_{min} = 2.8 \cdot 10^{-6} \, \mathrm{F/cm^2}$ for SU-2500 samples. For samples made of the SU-2500 material, the $C(\varphi)$ characteristics measured under several times repeated cycling of φ are given in Fig. 5.40 (curves 1–4). The curves show two peaks at 0 and $+0.45 \, \mathrm{V}$ with a relaxation time of less than 10^{-6} s (see curves 3 and 4).

The occurrence of maxima in the $C(\varphi)$ curves may be due to the reconstruction of the interface, starting from the hydride state of the surface to the hydroxide one. To reduce the effect of these processes, the capacitance measurements were carried out at reduced temperatures ($T = 250 \, \mathrm{K}$). The high capacitance value obtained from the "charging" curves $C = j(\mathrm{d}t/\mathrm{d}\varphi)$ and the absence of effects of adsorption of solution components and reconstruction on the graphite electrode surface suggest that the measured differential capacitance is completely defined by the SCL of the graphite electrodes. Additional evidence is the absence of frequency dispersion of the capacitance (see curves 1 and 2 in Fig. 5.40).

An estimation of the densities of electron and hole states in the allowed bands in the near-surface region of the graphite electrode indicates that their values near the band edges are close to the theoretical ones [153] and those measured in [154] for the bulk; they are in the range $(1–2) \cdot 10^{19} \, \mathrm{cm^{-3}}$. With further penetration of the Fermi level into the allowed bands in the process of polarization the values of $\mathrm{d}N/\mathrm{d}E$ on the surface were found to be lower than in the bulk [154].

81

Within the Kane-type dispersion approximation, the ratio of effective masses of the states density in the allowed bands in the graphite SCL was found from the experimental $C(\varphi)$ curves. That value is $m_{dlh}/m_{de} = 1.3–1.5$, which is in fair agreement with the bulk data [153].

Thus, it can be assumed that the electronic properties of a near-surface layer of graphite may well be treated within the SCL theory, generalized to the case of a nonquadratic dispersion law for allowed bands. The authors of [151] suggest that lower values of electron state densities in the SCL as compared with those in the bulk may be due either to formation of an energy gap in the SCL region resulting from the lifting of band degeneracy [which actually corresponds to the transition from a semimetal (gapless) state to a narrow-gap one] or to a decrease of the state density near the surface due to size quantization in a strong electric field.

5.4 FESE TECHNIQUE AS APPLIED TO STUDIES OF SURFACES OF METAL ELECTRODES AND HTSC MATERIALS

The FESE technique developed for studies of narrow-gap and zero-gap semiconductors and semimetals, which employs the notion of the electron state density near the Fermi level, may prove to be useful in investigations of electronic properties of metal electrodes and HTSC material surfaces. This can be exemplified by studies performed on mercury electrodes [155] and those made of HTSC materials, obtained on the basis of multicomponent metal-oxide compounds.

5.4.1 Hg

The FESE on Hg electrodes was first investigated in [108] (see Fig. 5.41).

The capacitance-voltage characteristics have been measured in the pulsed mode with pulse duration of the order of 10^{-6} s. The current-voltage characteristics were taken with the potentiostatic variation of the electrode potential for different rates of cycling. The Hg surface was subjected to a 10% HNO_3 solution followed by etching in strong oxidizers to remove the organic remainder. All measurements were made in electrolytes composed of 0.1 mol aqueous KNO_3 with the addition of $C_{10}H_{16}N_2O_8$ at pH = 8 and $C_6H_7O_{11}$ at pH = 6. Measurements were performed in the temperature range of 238–313 K. In all

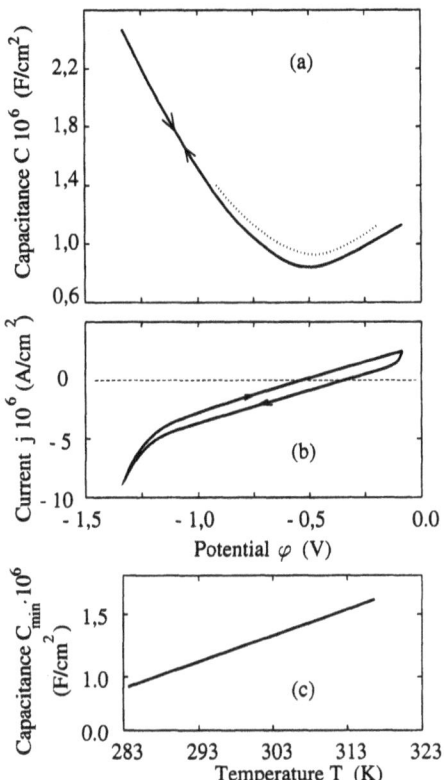

Figure 5.41 Capacitance-voltage curve (a), current-voltage curve (b), and dependence of the capacitance minimum on the temperature (c) for the Hg electrode. Dashed curve shows the influence of the light illumination.

cases, current-voltage curves corresponded to the charging curves for the geometric capacitance of the electric double layer at the phase boundary with $C = 20 \, \mu F/cm^2$. The capacitance-voltage curves obtained in these conditions proved to differ from the low-frequency ones obtained earlier for aqueous electrolytes [52, 156, 157]. These characteristics feature the U shape typical for narrow-gap semiconductors; a linear growth of C_{min} with increasing temperature and a reversible increase of capacitance under the electrode illumination with light are also characteristics of narrow-gap semiconductors.

The observed features in the capacitance-voltage characteristics may be accounted for as resulting from the U-like energy distribution of the electronic states densities in Hg near the border with the electrolyte. Yet they can also be explained by the presence of a forbidden gap in the distribution of states densities. The latter assumption seems to be more plausible, since it explains both the U-like character of the capacitance-voltage characteristics and the capacitance minimum rise with temperature and under illumination as well. Formation of a forbidden gap in the Hg near-surface layer may occur through the adsorption of $C_{10}H_{16}N_2O_8$ molecules on the Hg surface, capable, under appropriate circmustances, of producing a spatially ordered layer. In the case of a crystalline surface of semiconductors, the occurrence of such a layer on the electrode surface is a cause of the restoration of bulk atomic structure at its surface, provided the relevant structure parameters of this layer conform to those of the semiconductor crystal surface. For the Hg surface, which does not possess an ordered atomic structure and features, at the same time, a high atomic mobility, there can take place a kind of process of "epitaxy" of Hg atoms on the structurally ordered layer of adsorbed molecules. In this case, the minimum of free energy of the system is attained when the surface layer of Hg atoms reproduces the coordination order of the adsorbate. This quasi-two-dimensional Hg layer, whose quasi-period is dictated by the molecules, arrangement in the surface adsorbed layer, can be characterized as a semiconductor. It is known that the decreasing of the Hg density below $5\,g/cm^3$ leads to a metal–semiconductor transformation with $E_g > 0$ due to the splitting off of the p-like conduction band from the s-like valence band [158].

5.4.2   High-Temperature Superconductors

Reviews on the electrode properties of HTSC materials can be found elsewhere; see [159–162]. They discuss such issues as HTSC material synthesis in electrochemical cells and analysis of the potentialities of low-temperature electrochemical techniques for their study [161]. Electrolytes with a low freezing temperature which can be used for such studies are listed in Table 5.2.

Most of the electrophysical property studies of HTSC electrodes were performed on YBaCuO. The interest was largely due to the problem of degradation of this material. It was shown that its surface structure is a function of the electrolyte pH, exposure time, and polarization conditions. The structure of YBaCuO ceramics is severely altered when brought into contact with aqueous solutions of H_2SO_4 and Cu_2SO_4; the superconducting properties are changed as well. The properties were found to be stable in alkaline media [163–165]. The critical temperature for the transition to the superconducting state (T_c) was unchanged in an etchant of content $H_2O:HCl:H_3PO_4$

TABLE 5.2
Low-Temperature Electrolytes

T (K)	Electrolyte
88	Chloroethane, butyronitrile
115	Bromoethane, butyronitrile, isopentane, methylcyclopentane
128	Bromoethane, butyronitrile
155	69% propionitrile, 31% butyronitrile
170	Propionitrile, butyronitrile, ethanol
180	Methanol, dichloromethanol
192	Ammonia, isopropanol, dimethyl formamide

in the proportions 1:1:0.5, 1:1:1.5, and 1:1:2 [166]. Exposure to water resulted in a certain decrease of the value of T_c [162], whereas exposure to water vapor resulted in its growth [167]. Most organic solvents, such as acetone, methanol, ethanol, propanol, and hexane, do not, in fact, interact with $YBa_2Cu_3O_y$ [168].

Measurements of the surface photoelectromotive force (PEMF) on a YBaCuO and BiPbSrCaCuO film boundary with the electrolyte 0.1 mol KCl solution in ethylene glycol and the current-voltage characteristics of the contact at 300 K were carried out in [169]. The current-voltage characteristics were typical for semiconductor electrodes of p type. The optical width of the forbidden gap, found from the spectral distribution of the PEMF, was 1.35–1.36 eV for both compounds. This allows us to suppose that the photoelectric properties of HTSC electrodes are defined by the structural element Cu–O, which is common for both of the investigated materials.

Studies of properties of YBaCuO electrodes in aqueous electrolytes with employment of the FESE technique were carried out in [170]. These studies were performed on two sets of ceramic specimens of $YBa_2Cu_3O_6$ and their single crystals in 1 mol KOH-water solution at room temperature. The specimens from the first set were transformed into the superconducting state at $T_c = 92$ K. The specimens of the second set revealed no superconductivity upon heat treatment at 1200 °C. The field effect was revealed in both types of specimens in the range of $-0.8 < \varphi < 0$ V as reversible and reproducible dependencies of capacitance and current on the electrode potential, characteristic of p-type semiconductors (see Fig. 5.42).

The specimens from the first set featured, in the potential range of $-0.8 < \varphi < 0$ V, a current-voltage dependence characteristic of

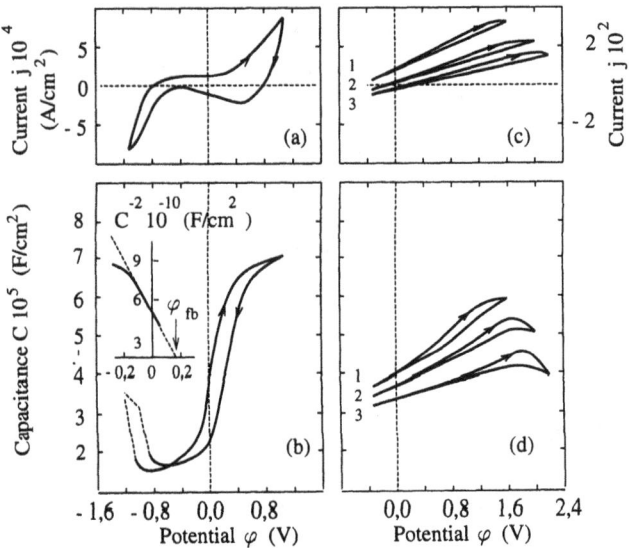

Figure 5.42 Current-voltage (a), (c) and capacitance-voltage (b), (d) characteristics of YBaCuO electrodes in 1 mol KOH aqueous electrolyte: characteristics (a), (b) are for specimens of the first set ($T_c = 92$ K); (c), (d) for specimens of the second set. In (b) the dashed line corresponds to the irreversible increase of the capacitance by cathodic polarization. In (c) and (d) curves 1, 2, and 3 correspond to different intervals of anodic potential changes.

a Schottky barrier (Fig. 5.42a). The capacitance-voltage character-istics (Fig. 5.42b) also followed a Schottky-Mott dependence and permitted evaluation of the hole concentration, which was found to be $p = 5 \cdot 10^{21}$ cm^{-3}. Juxtaposition of the value of $\varphi_{fb} = 0.17$ V with the value of the electrode potential measured in the absence of polarization ($\varphi_0 = 0.40$ V) showed that, in the unpolarized state, the YBa$_2$Cu$_3$O$_6$ surface is enriched with holes (upward band bending). Results of measurements for the specimens of the second set are presented in Figures 5.42c and 5.42d. In this case the potential φ_0 proved to be negative which corresponds to n-type conductivity of the near-surface region of the specimen. The increase of the range of electrode potentials resulted for both types of specimens in the occur-rence of irreversible regions in the $j(\varphi)$ and $C(\varphi)$ dependences, which is typical for processes of anodic oxidation.

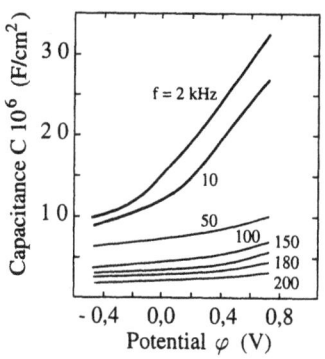

Figure 5.43 Variation in capacitance-voltage characteristics of $Y_1Ba_2Cu_3O_{6-7}$ electrodes (superdense ceramics) in KOH aqueous solution (pH = 12) with frequency of the probe signal.

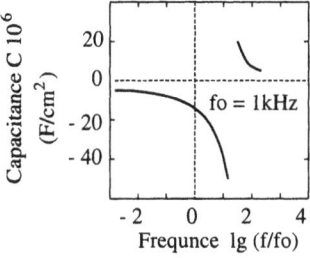

Figure 5.44 Frequency dependence of the capacitance.

Measurements of the capacitance-voltage curves versus frequency of the probe signal revealed their strong frequency dispersion in the region of anodic potentials (Fig. 5.43). The frequency dependence of the capacitance taken from Figure 5.43 at $\varphi = 0.8\,V$ is shown in Figure 5.44. The capacitance proves to be negative at frequencies below 10 kHz; which testifies to the inductance properties of the system and is characteristic of inductance-capacitance resonance. This means that the equivalent circuit for the HTSC material–electrolyte boundary can be represented by an *RCL* circuit, where the presence of the inductance can be explained due to the piezoelectric effect on the HTSC electrodes arising under their polarization.

Processes of Spatial and Temporal Self-Organization in the Semiconductor–Electrolyte System and Their Manifestation in the Field Effect

IF WE CONSIDER the problems concerned with self-organized processes in terms of thermodynamically open nonequilibrium systems that undergo nonlinear interactions, then a semiconductor surface that is brought into contact with an electrolyte offers considerable scope for the realization, observation, and investigation of such processes. Two types of subsystem can be distinguished in such a system: those of electrons and ions, which differ from one another in the charge relaxation time. The screening of charge in both of the subsystems is describable by the Poisson equation, in which the electron and ion densities exhibit a nonlinear dependence on the potential. Effects that can be interpreted as manifestations of self-organized phenomena are observable in atomic rearrangements of the SE interface. These occur in certain circumstances under semiconductor electrode polarization and adsorption on its surface of a number of molecules from the electrolyte. These effects are revealed as steplike changes of FESE characteristics and have features typical of phase transitions, resulting in the formation of ordered quasi-two-dimensional electronic and atomic layers on the semiconductor surface.

6.1   PROCESSES UNDER POLARIZATION

An example of formation of an ordered structure in a SE system is the germanium electrode surface in neutral solutions, after anodic polarization, when the $C(\varphi)$ relationship acquires the form characteristic

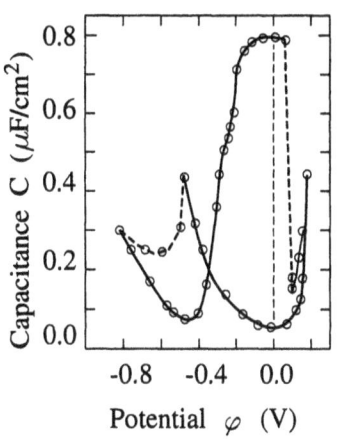

Figure 6.1 Dependence of the differential capacitance on the electrode potential for p-Ge ($\rho = 3\,\Omega \cdot$ cm) in 48% HF aqueous solution.

for the dependence of capacitance on V_s. This gives evidence in favor of a high enough degree of phase boundary ordering (see Chapter 3) and attests that the ionic (atomic) processes occurring on this type of surface result, at certain values of pH, in the formation of spatially ordered structures. Such surfaces are quite stable within a certain interval of potentials, whereas the U-shaped $C(\varphi)$ curve corresponds to the minimum of the electric energy.

It is well known that, in the process of rearrangement, the atomic structures usually pass through a number of quasistable states. This is a characteristic feature of evolution processes in general, and underlies their dynamics [45]. An example of the realization of such steady states in the SE system is the behavior of germanium in a 48% solution of HF [3] (see Fig. 6.1). The presence of two reproducible U-shaped $C(\varphi)$ curves shows that, by polarizing the germanium electrode, one can obtain two different quasistable states. The magnitudes of the capacitance at the minimum in the $C(\varphi)$ curve are close to the theoretical value of the capacitance of the SCL for germanium (with the roughness coefficient taken into account), which is indicative of the ordered character of the interface structure. The authors of [3] suggested that these two states differ from one another by the amount of adsorbed oxygen. The transition from one quasistable state to another, accompanied by the displacement of the $C(\varphi)$ curve by 0.6 V, was ascribed to a change in the interfacial potential drop on oxidation and reduction of the germanium surface. Yet, taking into account that $C_H \sim 20\,\mu\mathrm{F/cm^2}$, the corresponding change in charge will

89

Figure 6.2 Variations in the capacitance (a), current (b), and conductance (c) with the electrode potential for p-Ge in 1 mol KCl aqueous solution, measured in the potentiostatic condition.

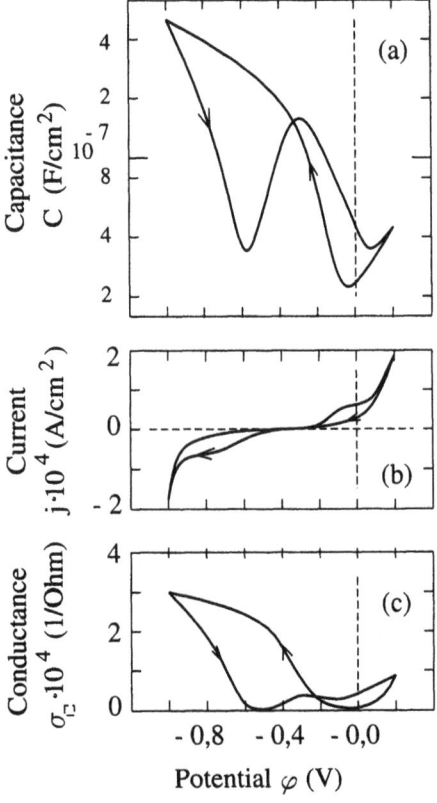

take a value of the order of 10^{-5} Cb/cm^2 which is almost ten times less than that required for reduction of the adsorbed oxygen.

The transition of the germanium-electrolyte system through successive quasistable states can also be exemplified by the germanium behavior in neutral and acidic aqueous solutions at the transition from the anode region of polarization to the cathode one and in reverse order [171–173]. Such a transition is characterized by the occurrence of two stable minima in the dependence of the capacitance and conductivity on the electrode potential, corresponding to the two quasistable states of the germanium electrode surface which switch back and forth from one to the other (see Fig. 6.2). These states correspond to a certain degree of order with respect to the atom arrangement at the interface. This is attested by the U-shaped character of the $C(\varphi)$ curve which coincides with the theoretical dependence of C_{sc} on V_s, as well as by low values (about 10 cm/s) of surface

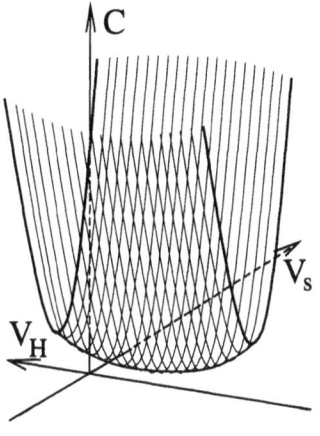

Figure 6.3 $C(V_s, V_H)$ surface for germanium electrode–aqueous electrolyte system.

recombination velocity, independent of the electrode potential. At the same time, in the transition region of potentials (in the range between the two minima) high surface recombination velocities are observed, indicating a disordering of the surface. It might be suggested that the region in question is related to the region in which there is observed a transition from one ordered state of the interface to another one through disorder.

In order to explain the observed features of the FESE, in particular those revealed in the $C(\varphi)$ characteristics, it is expedient to consider the capacitance of the system as a function of the two variables V_s and V_H. Here, it seems reasonable to treat the problem by introducing a three-dimensional space with C, V_s, and V_H as coordinates. In these coordinates, different states of the SE system are represented as a set of points on the $C(V_s, V_H)$ surface. Consider, for example, the germanium electrode–aqueous electrolyte system. In this case, the $C(V_s, V_H)$ surfaces are trough shaped (see Fig. 6.3). For intrinsic germanium, one of the walls of the "trough" corresponds to electron accumulation, while the other one relates to hole accumulation, with the cross section of the surface at $V_H = \text{const}$ defining by the charge changes in the electron subsystem of the semiconductor SCL and corresponding to the absence of atomic rearrangements at the boundary. It is assumed herewith that the conditions of ideal polarizability for the semiconductor electrode are realized, and the measured capacitance can be interpreted within the quasi-equilibrium FESE. The cross sections at $V_s = \text{const}$ and for different values of V_H are defined by alterations in the ion subsystem

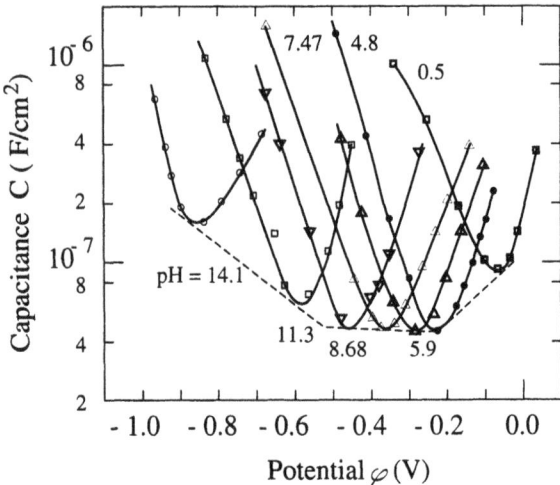

Figure 6.4 Capacitance-voltage characteristics of the germanium electrode for different pH electrolyte.

and reflect the atomic (ionic) rearrangement of the semiconductor surface and of the SE interface under the condition that the total charge in the system is conserved. The cross sections in question can be deduced from measurements, in quasi-equilibrium conditions, of the minimal values of $\sigma(\varphi)$ and $C(\varphi)$ curves versus the electrolyte pH, which are U shaped for germanium electrodes in aqueous electrolytes[1] [171] (see Fig. 6.4).

A clear idea of the dynamics of the change of the semiconductor-electrolyte system states under polarization and its manifestations in the FESE become evident using the concept of potentiodynamics trajectories, which reflect simultaneously alterations of potentials occurring in the SCL (V_s) as well as those in the Helmoltz layer (V_H) taking account of their interrelation [174, 175]. The actual appearance of these trajectories is a function of the relation between the charge relaxation rates in the electron and ion subsystems. From this it follows that, in the presence of atomic reconstructions at the interface and semiconductor surface, it may be that the $C(\varphi)$ characteristic will fully occupy the position at only one wall of the trough. The occurrence of a minimum in this situation is not necessarily associated with

---

[1] Analogous dependences of $C(\varphi)$ curves on the electrolyte pH value were also observed for InAs, InSb, and (CdHg)Te (see Figs. 5.14, 5.18, and 5.24, Chapter 5).

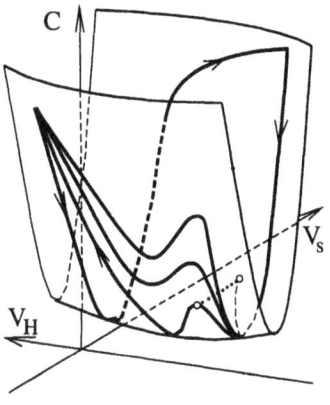

Figure 6.5 Family of potentiodynamic trajectories for germanium electrode in aqueous solution.

any change in the sign of the charge carriers near the semiconductor surface, as is usually assumed in conventional approaches to interpretation of $C(\varphi)$ characteristics within the quasi-equilibrium FESE. This means that the $C(\varphi)$ characteristics may not be used for describing the processes in the SCL only on the ground that they are U shaped, and supplementary analysis of the situation is necessary.

Figure 6.5 shows a family of potentiodynamic trajectories, illustrating the transition from an anode-polarized germanium surface to a cathode-polarized one [172].

The appearance of minima in the $C(\varphi)$ curve at potential values exceeding the value of φ at the bottom of the trough, resulting from the change in the sign of charge carriers, would mean a steplike transition of the state of the SE system from the "hole" wall of the trough to the "electron" one (see the dotted line in Fig. 6.5). But such a transition would correspond to a breakup of continuity of the trajectory, and, correspondingly, the reversal of the polarization current direction, which is not observed in experiments. Hence, with the polarization current direction unchanged, the minimum in the $C(\varphi)$ curve observed at potentials above the value corresponding to the trough bottom is not a consequence of the reversal of the charge carrier sign at the semiconductor surface, but is the result of atomic reconstruction of the germanium surface when one passes from anodic to cathodic polarization.

At the present time, the detailed character of the reconstruction is not clearly understood, nor to what atomic (ionic) configurations the observed stable surface states correspond. In a number of papers (see, for example, [172, 173]) it is assumed that the two quasistable states of the germanium surface which arise as a result of the transition from anodic to cathodic polarization are associated with hydroxide and hydrate coatings on the surface. It was assumed herewith that, as one passes from the hydroxide to the hydrate state and back, there appears in the intermediate region a considerable amount of non-saturated (dangling) bonds on the surface, which, in fact, leads to a sharp increase of surface recombination velocity [176, 177]. Yet the stepwise change of the electrode potential observed at this transition does not, as a rule, correspond to the difference between the Ge-H molecule dipole moment and that of Ge-OH [171], which makes the explanation questionable. Without going into the details of the process, which are somewhat unclear, it can be stated that there exists reproducible experimental evidence in favor of the existence, in the germanium–aqueous electrolyte system, of two capacitance and conductance minima with specific hysteresis of the $C(\varphi)$ and $\sigma(\varphi)$ curves exhibited on continuous scanning of φ in the potentio-dynamic regime. This indicates the fact that there exist two stable states of the interface, reproducible in n- and p-type germanium for different crystallographic orientations of the surface. From this it can be concluded that, for a given system, there exists at least one bistable state which can be revealed in FESE. Correspondingly, this means that self-organized phenomena may arise in this system [45].

6.2 Processes under Adsorption

Another manifestation of self-organized effects in the FESE is found on a semiconductor electrode surface on the adsorption of complex organic molecules with an a priori known stereometric configuration, which were introduced into the electrolyte and had different confor-mation states at various different electrolyte pH values. The molecules were chosen proceeding from their structure in the solution and their characteristic dimensions, with reference to the compatibility of the latter with the semiconductor lattice parameters. Figure 6.6 demon-strates the $C(\varphi)$ characteristics for the system germanium [p type, $\rho = 25\,\Omega \cdot \text{cm}$, (100)] – 1 mol KCl (pH = 6). They correspond to various intervals of φ in the cathode region of polarization. The analysis of these curves taken in the anode-cathode direction was performed up to values of φ well below the potential value relating to cathode isola-tion of hydrogen (Fig. 6.6a). It points to the presence of surface states

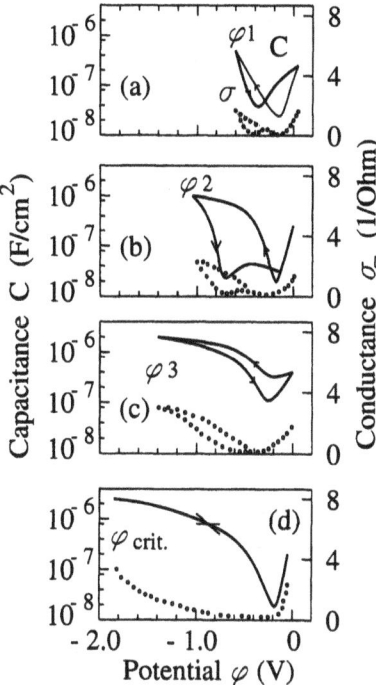

Figure 6.6 Variations in the capacitance and surface conductivity with the electrode potential for p-Ge electrode in 1 mol KCl aqueous solution: (a)–(c) pure KCl solution, various intervals in the cathode region of polarization; (d) KCl solution with glucose.

on the germanium surface which are unevenly distributed through the band gap. Their density N_{ss} increases from 10^{10} cm^{-2} near the midgap up to more than 10^{12} cm^{-2} near the conduction band bottom. Such a density distribution is indicative of the heterogeneity of the system under consideration [64, 66].

By extending the range of the electrode potential variation in the region of cathode values, an additional cathode minimum was revealed (see Fig. 6.6b), which is usually associated with change of the surface coating from hydroxide to hydride. With further extension of the interval of electrode potential variations, the capacitance at the cathode minimum increased, which might be the result of an increase in the number of surface states on cathode-induced enrichment of germanium with hydrogen. At the same time, an increase of capacitance was observed at the anode maximum (see Fig. 6.6c), which points to a further disorderning of the germanium surface. With introduction of a glucose solution into the electrolyte the $C(\varphi)$ curve behavior was drastically changed, when, during the polarization process, φ reached a certain critical value $\varphi_{crit} = -1.8$ V (see Fig. 6.6d). Here, the $C(\varphi)$ curve changed abruptly: (a) the hysteresis

95

was observed to disappear; (b) narrowing of the curves occurred; (c) they acquired a stable form, which did not undergo changes during repeated cycles of φ variation, and corresponded, within the whole interval of surface potentials, to the theoretical one for the germanium SCL. Moreover, the capacitance value at the minimum coincided with the theoretical one for the minimum of the SCL capacitance (without taking account of the roughness). Note that, if we make narrow the range within which φ is varied in the cathode polarization ($\varphi < \varphi_{crit}$), then, at the following cyclic scanning, the curves coincide and retain a stable form during repeated cycling of φ, whereas the capacitance at the minimum retains its value.

The results obtained demonstrate that, on achieving φ_{crit}, there occurs a stepwise transition from the disordered state to one with a high degree of order on the SE interface. This order has a steady character, which further is retained within the whole interval of φ. It is seemingly associated with the formation of an ordered layer of adsorbed molecules with a high degree of homogeneity over the surface. That fact that this transition is stimulated by external perturbation and has a steplike character is in favor of the following: its character corresponds to a disorder-order transition of the self-organization type [45].

Another example of the formation of a stable germanium-electrolyte interface featuring a high degree of ordering is given in Figure 6.7. In this case, molecules of EDTA, ammonia, and NH_2CSNH_2 were added to a Na_2SO_4 solution. Following a certain cycle of φ changes, there was observed the formation of a stable boundary, which showed an ordered structure, as evidenced by (a) a good agreement between the interface $C(\varphi)$ curve shape and the one that follows from the theoretical $C_{sc}(V_s)$ relationship; (b) the coincidence of the values of C_{exp} and C_{theor} at the minima of the $C(\varphi)$ and $C_{sc}(V_s)$ curves, respectively.

Figures 6.6d and 6.7 (curve 1) show the $C(\varphi)$ characteristics of a germanium electrode in a solution of pH chosen such that the structure of complex organic molecules in the solution is most favorable for their ordered arrangement on the germanium surface when adsorbed. Variation of pH with respect to the chosen optimal conditions resulted in less favorable conditions for such an arrangement of the molecules. In this case formation of an ordered surface structure either was not observed at all, or occurred during a long period of time, and was realized only through the subsequent several cycles of electrode

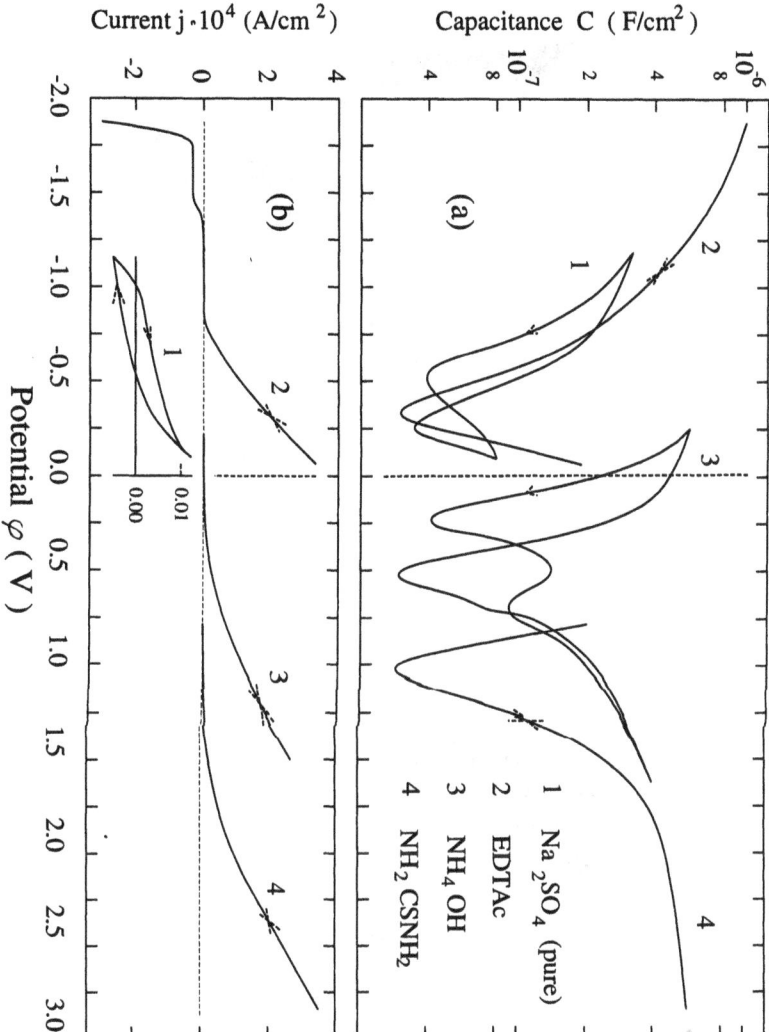

Figure 6.7 Capacitance-voltage (a) and current-voltage (b) characteristics of Ge electrode [p type, $\rho = 25\,\Omega \cdot$ cm, orientation (111)] in different aqueous electrolytes.

potential variations. A characteristic example of the formation of a stable ordered boundary is represented by EDTA molecule adsorption from a solution of pH = 5, which is given by a series of potentiodynamic characteristics as in Figure 6.8. In the solution of pH = 8, on the addition of EDTA molecules, formation of the stable state took

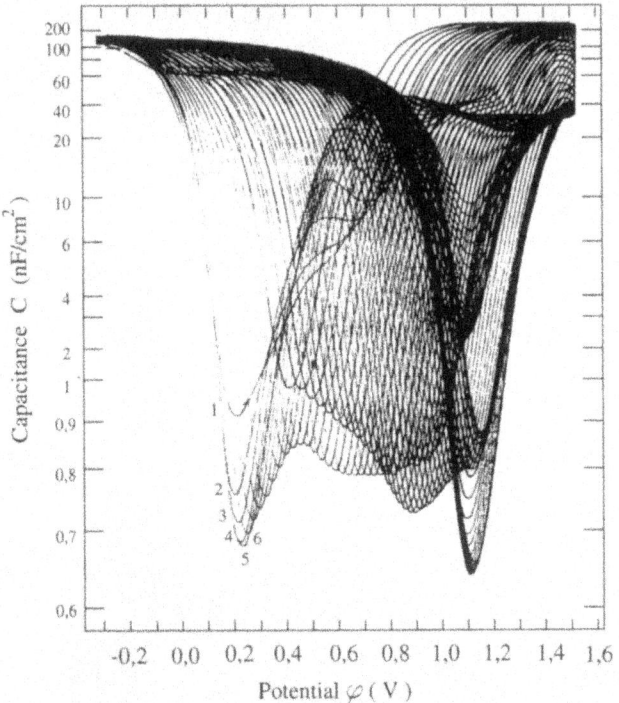

Figure 6.8 Kinetics of capacitance-voltage characteristic change for Ge electrode [p type, $\rho = 25\,\Omega \cdot$ cm, orientation (111)] in aqueous solution of EDTA (pH $= 5$). The early passages to the anode from the cathode are numbered as $1, \ldots, 6$.

place within one cycle and was further reproduced completely from cycle to cycle (see Fig. 6.7, curve 1). In contrast, in the case of pH $= 5$, a stable state could only be realized after a large number of polarization cycles. The presence of a great number of cycles is likely to reflect the process of "finding out" by the system of this correlation between the surface and molecular orientations (by the proper "choice" from differently oriented molecules), when their orientation would correspond to their most stable and ordered arrangement on the surface.

Figure 6.9 represents the character of temporal changes of the boundary capacitance, corresponding to the equilibrium potential (in the absence of an external polarization) in a KCl solution of pH $= 5$ prior to and following the introduction of EDTA molecules into the solution. In contrast to the $C(t)$ dependence in a pure KCl solution, which has the appearance of a smooth curve (see Fig. 6.9, curve 1),

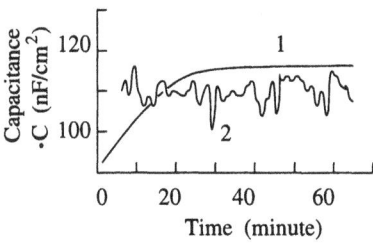

Figure 6.9 Variations in the capacitance with time for p-Ge electrode in the unpolarized state in pure 1 mol KCl aqueous solution (1) and in KCl aqueous solution after addition of EDTA (2).

after introducing the EDTA molecules, the dependence has a noise-like character (see Fig. 6.9, curve 2). Autocorrelational analysis of this dependence reveals the presence in it of quasiperiodic oscillations with periods of 2 and 8 min, which, according to synergistic notions, could bear witness to the occurrence of spatial-temporal self-organization [45]. Variations in pH within 5–6 brought about alterations in the amplitude and frequency of the quasi-periodic oscillations. The latter were removed at pH = 8. The physical origin of charge oscillations at the interface is, probably, the fluctuations of the orientations of adsorbed molecules having different conformational states. In view of the known strong influence of pH on the conformational properties of EDTA molecules [178], it can be suggested that the value of the pH is a governing parameter for the system [175].

As in germanium, features can be observed on surfaces of other semiconductors under an appropriate choice of the electrolyte pH and the type of adsorbed molecule. Examples are presented in Figure 6.10 for InSb and (CdHg)Te in contact with a 15-Crown-5 benzene solution. Similar results were obtained with semiconductors such as GaSb, HgTe, and (CdHg)Te in contact with aqueous electrolytes. It is worth noting that in these cases a correlation between experimental $C(\varphi)$ and theoretical $C_{sc}(V_s)$ characteristics was found and no hysteresis was observed. Also remarkable is the fact that several stable U-shaped regions were observed in (CdHg)Te and GaSb, within which the $C(\varphi)$ characteristics were of reproducible character under many times cycling of the electrode potential (see Figs. 6.11 and 6.12).

The results outlined above point to some novel possibilities of controlled change of electrophysical properties of semiconductor surfaces, and, in particular, to new possibilities of their production, with practically no surface states throughout the semiconductor band gap. Formation on such surfaces of thin coatings with insulating properties, making use of adsorption of particles of high-molecular-weight compounds, enables a significant extension of the range of

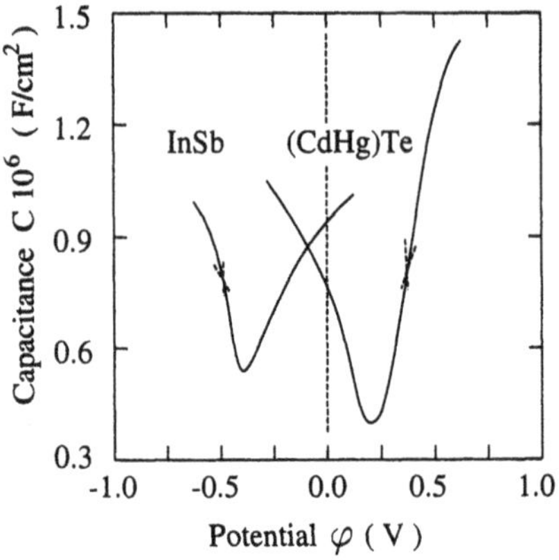

Figure 6.10 Capacitance-voltage characteristics of InSb and $Cd_{0.2}$ $Hg_{0.8}Te$ electrodes in 15-Crown-5 benzene solution illustrating the dependence of potential change in double electric layer on the type of semiconductor.

Figure 6.11 Capacitance-voltage (a) and current-voltage (b) characteristics of (CdHg)Te electrode in aqueous electrolyte.

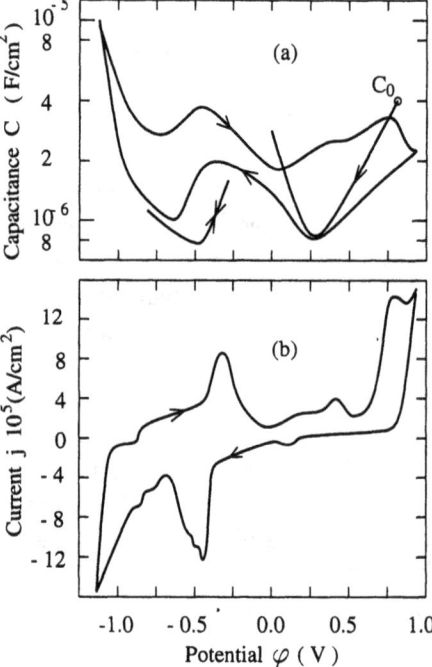

ideal polarizability of semiconductor electrodes. This, in its turn, opens new perspectives for studying electrophysical characteristics of surface and near-surface layers by the FESE. Moreover, the formation of thin layers by means of adsorption of appropriate molecules that passivate the semiconductor surface with a low density of surface electronic states and provide stability to various different external

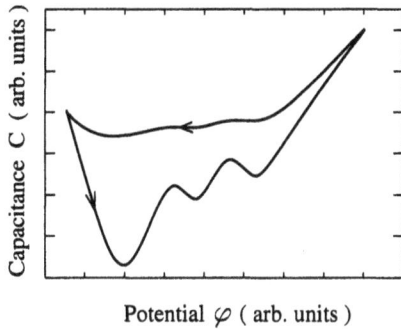

Figure 6.12 The type of capacitance-voltage characteristics of GaSb electrode in KCl + EDTA aqueous solution.

perturbations may be important for semiconductor device production. In particular, this may be useful for A^3B^5 compounds, since it enables, prevention of Fermi level pinning on the semiconductor surface [62] (see Chapter 5).

6.3 QUASI-PERIODIC OSCILLATIONS IN
THE CARBON FIBER–ELECTROLYTE SYSTEM

The field-effect method in semiconductor-electrolyte systems was used for investigation of the electrophysical properties of the carbon fiber and electrolyte interface. The Carbon fiber was a sample of threadlike form with a diameter about $10\,\mu$m and had a length of 2.5 cm. The ohmic contact to the fiber was obtained with eutectic InGa or carbon soot. The area of contact was protected from the electrolyte by special wax. The working electrolyte contained solutions of KCl, EDTA, and NH_4^+. That ensured a wide range of polarization of the system without escape of hydrogen by cathodic or oxygen by anodic polarization. In experiments the current (I) through the carbon fiber–electrolyte interface, the differential capacitance (C), and the electrode potential (V) of a sample were measured in a wide range of frequencies ($0 < f < 100$ Hz) by sample polarization in potentiostatic conditions with the time constant stabilization above 3 s.

The dependence of the electrode potential on the time of sample polarization is presented in Figure 6.13. It is revealed that achieving the critical value of the electrode potential in the cathodic region ($\varphi_c \approx -7.5$ V) allows for quasi-periodic oscillations with time constant less than the time constant of potential stabilization. It is characteristic that attempting to increase the cathodic potential beyond its

101

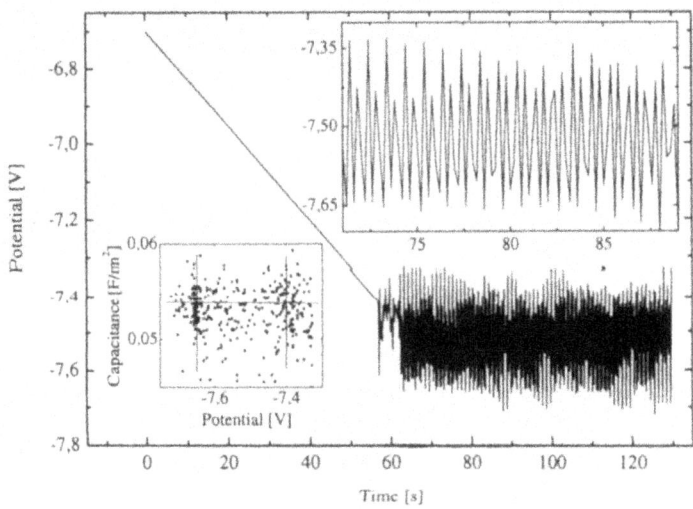

Figure 6.13 Complicated behavior of the electrophysical parameters of the carbon microelectrode–electrolyte system during a polarization process.

critical value ($|\varphi| > |\varphi_c|$) results in increase of the frequency of observable oscillations. Here, the constant component of the electrode potential remains practically unchanged ($|\varphi| \simeq |\varphi_c|$). When $|\varphi| < |\varphi_c|$ the oscillations disappear. The dependence of the variable component of the electrode potential near $|\varphi| = |\varphi_c|$ is given in the upper inset of Figure 6.13.

The dependence of the capacitance on the potential inside the region of oscillation (see the bottom inset of Fig. 6.13) may be characterized by at least two fields with denser condensation of points. This shows that the carbon fiber–electrolyte system has apparently not less than two states of charge accumulation at the carbon fiber–electrolyte interface.

The dependence of current on the potential (Fig. 6.14a) is represented by the closed curves. If the points of these curves are connected by straight lines according to the sequence of their appearance in the process of measurement (Fig. 6.14b) the dependence looks like the trajectory wound on a toroid. In the process of increase of the number of measured points the winding toroid becomes more and more dense. This result is typical for nonlinear dynamic systems characterized as strange attractors [45].

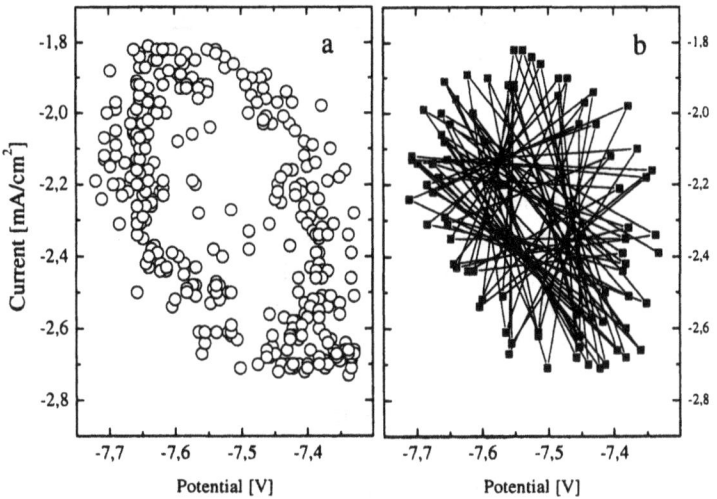

Figure 6.14 The current-voltage characteristic of the carbon microelectrode–electrolyte system under a fixed cathodic polarization.

The physical mechanism of the observable oscillations of the measured parameters is apparently connected with quasi-periodic infringements of the adsorption-desorption balance on the homogeneous surface of a fiber with simultaneous size quantization of electrons in the space charge region during polarization.

Oscillations defined by the given mechanism include the following intervals of time corresponding to spatially divided charging processes, namely,

- time of accumulation of charge in the electrode-electrolyte interface with simultaneous polarization of the SCL of the semiconductor;
- time of capture of charge from SCL of the semiconductor carbon fiber on the discrete level of an $EDTA^{4-}$ ion in surface layers with simultaneous depolarization of the semiconductor SCL;
- time of charge injection from the surface layers ($EDTA^0$) into the electrolyte bulk with simultaneous polarization of the semiconductor SCL but already in the next cycle of oscillation.

It is necessary to note that the first and third specified intervals of time correspond to the same mechanism for two consecutive periods of oscillation. At this stage the time of relaxation is given by diffusion of $EDTA^0$ molecules into the bulk of electrolyte with their replacement by $EDTA^{4-}$ ions. This time appears much longer than the time of electron capture occurring during the preceding stage. Within the framework of the offered mechanism the oscillations will exist if the electron concentration in the SCL of the semiconductor (at the moment of capture of a charge on the $EDTA^{4-}$ ion level) is approximately equal to the concentration of electronic traps created by $EDTA^{4-}$ ions in the surface layer. Otherwise there will be a dynamic balance of charges in the system. The experiment (see the bottom inset of Fig. 6.13) gives the difference between the concentrations of charges for two conditions corresponding to φ_1 and φ_2: $\Delta N = (C\Delta\varphi)/q = 1.6 \cdot 10^{12} \, \text{cm}^{-2}$, where $\Delta\varphi = \varphi_2 - \varphi_1$.

For carbon, with electron effective mass in the conductivity band $m_e = 0.03 m_0$, with the assumption of an electrical quantum limit (the case of filling of one quantum subband) the concentration of 2D electrons (N_{2D}) may be estimated from the correlation $\Delta N_{2D} = (m_e^2 \Delta V_s)/(\pi^2 2q) = 1.8 \cdot 10^{12} \, \text{cm}^{-2}$, where ΔV_s is the variation of the surface potential, whose value can be appreciated from the bottom inset of Fig. 6.13 using the formula $\Delta V_s = (\varphi_{(C=5\,\mu F/cm^2)} - \varphi_{C_{min}}) = 0.15 \, \text{V}$. The coincidence of values of the concentrations $\Delta N = \Delta N_{2D}$ confirms the legitimacy of the assumption about the mechanism of the oscillation appearance. Additionally, the assumptions made explain naturally the increase of the frequency of oscillations on increase of the integrated current through the carbon fiber–electrolyte interface.

It is evident that processes of a similar nature may be observed for different kinds of fine-carbon materials. Figure 6.15 exemplifies two types of such materials which are synthesized in an explosion-generated pulse camera (Pulse-M). The essential volume of each

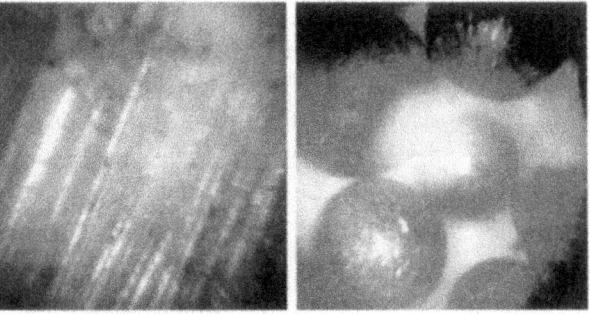

Figure 6.15 (a) Samples of fiberlike oblong carbon around 3–6 μm in cross-section; (b) carbon spheres 5–15 μm in diameter.

particle of these materials is the SCL region (see Section 5.3.3) so that the material as a whole has to have field-controlled properties.

6.4 BELOUSOV-ZHABOTINSKY REACTION

The field effect in a semiconductor-electrolyte interface is commonly described as the consequence of charge change in the space charge layer of the semiconductor. That change is the consequence of the change of ionic charge on a semiconductor surface following external voltage application (polarization of the semiconductor). However, the change of charge in the ionic part of a double layer on a semiconductor-electrolyte interface may be the consequence of other different processes (reactions) that occur in the electrolyte. In this case changes of charge in the semiconductor-electrolyte interface caused by these processes may be determined by measuring the capacitance, conductivity, and electrode potential of the semiconductor surface. This measurement may serve as a method to reveal and investigate these processes and allow their characteristics to be determined.

An example of these possibilities is the research into capacitance and electrode potential during the Belousov-Zhabotinsky reaction carried out by the authors of this monograph for the first time. This oscillatory reaction is well known since the time of its discovery [45]. It takes place in solutions of certain compositions and consists in periodic change of the solution color with time until one of the components of the solution is used up. Recently this reaction has evoked new interest because it may be considered as a manifestation of self-organization effects and may serve as a method for simulation and study of synergy processes in solid electrode–electrolyte interfaces.

There are many different combinations of solutions and their components in which this reaction may take place. The periods of oscillation and the solution color may be different. We investigate Belousov-Zhabotinsky's oscillatory reaction with $CH_2(COOH)_2$, $KMnO_4$, H_2SO_4, and KBr in a stagnant cell. The charge-sensitive Ge electrode was laterally vibrating with frequency 20 Hz. The reaction was realized accordingly to the following scheme [45] (Table 6.1).

Reaction 1 of Mn oxidation is carried out first. Reactions 1a and 1b are parallel. The Mn^{2+} formed reacts in accordance with reaction 2. The bromide reacts with bromate by reaction 3 followed by

TABLE 6.1
Belousov-Zhabotinsky Oscillatory Reaction

Phase	Reaction
1	$MnO_4^- + 8H^+ + 5e = Mn^{2+} + 4H_2O$
1a	$CH_2(COOH)_2 + KMnO_4 + H_2SO_4 = CO_2 + H_2O + MnSO_4 + K_2SO_4$
1b	$CH_2(COOH)_2 + KBrO_3 = CO_2 + H_2O + KBr$
2	$Mn^{2+} + BrO_3^- + 6H^+ = Mn^{3+} + Br^- + 3H_2O$
3	$Br^- + BrO_3^- = BrO_2^- + BrO^-$
3a	$BrO^- + Br^- + 3H^+ = Br_2 + H_2O$
3b	$BrO_2^- + 3Br^- + 4H^+ = 2Br_2 + H_2O$
4	$Br_2 + CH_2(COOH)_2 = CHBr(COOH)_2 + H^+ + Br^-$
5	$CH_2(COOH)_2 + Mn^{3+} = Mn^{2+} + CO_2 + H_2O$
5a	$CHBr(COOH)_2 + Mn^{3+} = Mn^{2+} + CO_2 + H_2O + Br^-$
6	$Br_2 + Mn^{2+} + e = 2Br^- + Mn^{3+}$

processes 3a and 3b. Free bromine consumption occurs in its reaction with malonic acid in reaction 4. Reaction 5 proceeds in parallel; it is identified by carbon dioxide bubbles escaping. After sufficient Br^- has been produced (reactions 2 and 4), bromide and bromate interaction begins. This interaction is followed by extraction of a portion of free bromine, which occurs spontaneously and leads to solution color change. At the same time the bromine that was extracted takes part in producing Mn^{3+} ions through reaction 6. After disappearance of free bromine and Mn^{2+} ions some malonic acid, bromine ions, and Mn^{3+} ions are still present in the reaction mixture. The reactions described above are repeated again and again until the moment that at least one component of the reaction mixture is used up. The repeating of these processes leads to quasi-periodic changes of the reaction mixture color.

In measurements the reaction mixture components were dissolved in water and then put into an electrochemical cell. The differential capacity and electrode potential of the semiconductor-electrolyte system were measured by use of two impulse methods. The electrode potential was measured relative to the Pt electrode using a high-input resistance voltmeter. The kinetics of the differential capacity and electrode potential were measured.

During the reaction the solution color variation and also that of the electrical parameters connected with bromine concentration

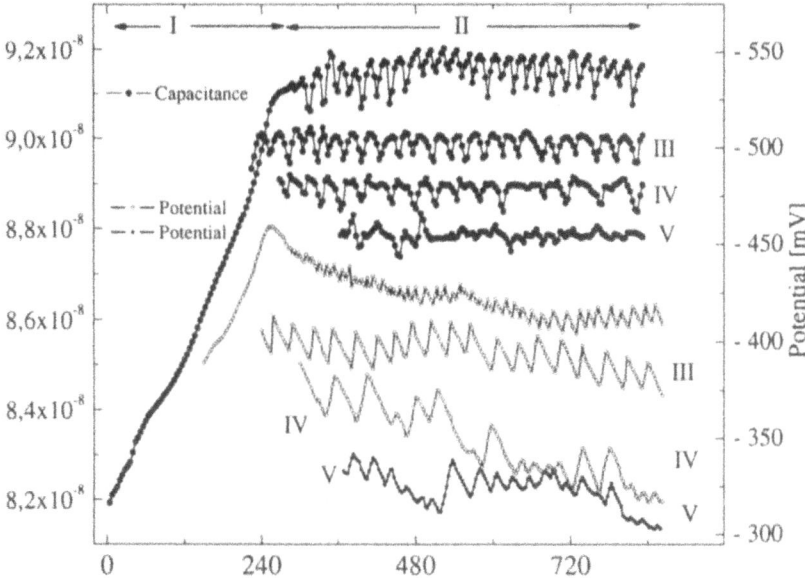

Figure 6.16 Typical instances of differential capacitance and electrode potential time characteristics.

fluctuations in oscillatory reaction were observed. It was found that the electrophysical parameters of the Ge electrode were correlated with the Br concentration swings.

Five regions with different kinds of differential capacitance variations can be distinguished (see Fig. 6.16). In the first region the capacitance and the voltage grow monotonically with time. This reflects the formation of new components necessary for the initiation of the autocatalytic reaction. The second region corresponds to the beginning of the oscillations. In the third region we observe stable oscillations. Here the concentration of the solution components can change only slightly. The fourth region corresponds to the depletion of the solution components in the closed cell. The color of the solution does not change, in contrast to the differential capacitance and the voltage, which continue to swing aperiodically and then decay in the fifth region.

It is important to note that we observe a pulselike change of the charge in the space charge layer of the semiconductor whereas the interface polarization is changing smoothly for each period of the oscillatory reaction. This shows that the change of the electrode

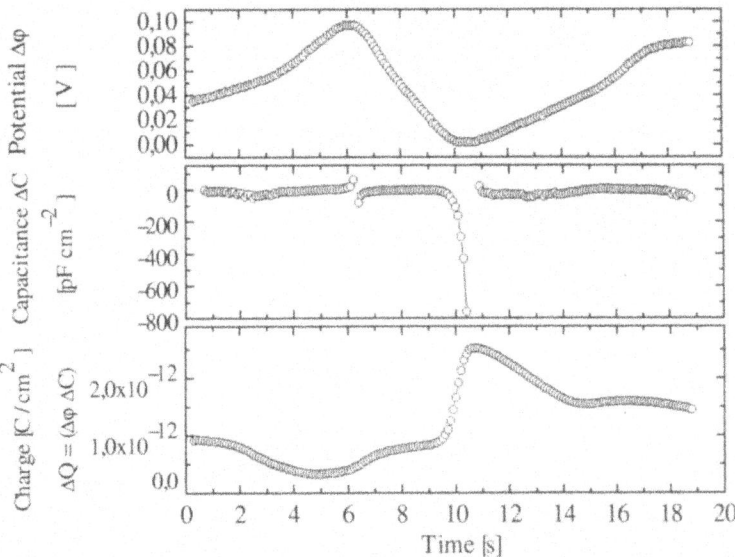

Figure 6.17 One of a series of experimentally measured oscillations of capacitance and potential and calculated charge–time characteristics.

voltage and the capacitance of the semiconductor-electrolyte system occurs during the oscillatory chemical reaction at the electrode-electrolyte interface.

The investigations carried out showed that the changes of electro-physical properties of the Ge electrode surface during the Belousov-Zhabotinsky oscillatory reaction give more details about the processes causing this reaction. It is obvious that this method of investigation of the processes that take place in solution in a number of cases may be very effective for determination of the mechanisms of these processes.

In addition to the Belousov-Zhabotinsky reaction many other reactions may take place in liquid surroundings, which lead to charge change on the semiconductor surface and may be revealed and investigated by this method. This use of semiconductor electrodes may be especially interesting for investigation of biological processes, in particular, for processes in biomembrane boundaries which are in contact with the surrounding biological liquid environment. A confirmation of this thought is the form of the electric signals from the interface obtained during the chemical oscillatory reaction, which look similar to electrophysiological signals from living systems (see Fig. 6.17).

Size Quantization in the Semiconductor-Electrolyte System

Recently there is growing interest in the quantum electronic effects which are realized in samples with geometrical sizes comparable with the electron wavelength (L_0) [179–192]. In this regard the most important result is the discovery of the weak electron localization effect and the quantum Hall effect [181]. Use of quantum effects led to the creation of electronics in which systems with two-dimensional, one-dimensional (quantum wires), and zero-dimensional (quantum dots) electron gases appear as the individual functional elements. In this chapter we consider some questions concerning the theory of two- and one-dimensional electronic systems and experimental manifestations of low-dimensional quantization in semiconductor-electrolyte interfaces. In the concluding part of this chapter we consider the possibilities of low-dimensional quantum structures formed in the (CdHg)Te-electrolyte system and investigations of their electronic properties on the basis of the field effect in the semiconductor-electrolyte interface.

7.1 Two-Dimensional Quantum Systems: Theory

When an electric field is applied perpendicularly to the surface of the semiconductor, a space charge layer is induced in the subsurface layer [8]. In the frame of the one-particle Hartree approximation the distribution of charge density is defined from self-consistent solution of the Poisson and Schrödinger equations [66]. In InSb at room temperature it is necessary to take into account simultaneously both types of charge carriers in the whole allowed range of energies. Therefore the calculation is based on the model of a "semi-infinite crystal with a second fictitious border" [190] with modifications implying the specific character of Kane semiconductors.

109

The bulk dispersion law of Kane semiconductors is described by the formula [116]

$$\mathbf{k}^2(E) = \frac{1}{P^2} \frac{(E - E_0)(E - E_0 + E_g)(E - E_0 + E_g + \Delta)}{(E - E_0 + E_g + \frac{2}{3}\Delta)}, \qquad (7.1)$$

where $\mathbf{k}$ is the wave vector, P is the matrix element accounting for the interaction between the conductance and valence bands, $E_0 = \hbar^2 k^2/(2m_0)$ is the energy of a free electron with mass m_0, E_g is the energy gap of the semiconductor, and Δ is the energy of spin-orbital splitting of the valence band.

In the general case the density of charge carriers in the bulk of the semiconductor is determined by Fermi-Dirac statistics

$$\rho = \frac{2}{(2\pi)^3} \int f_0(\mathbf{k}) \, d\mathbf{k},$$

$$f_0(\mathbf{k}) = \frac{1}{1 + \exp\{[E(\mathbf{k}) - E_F]/(k_0 T)\}} \qquad (7.2)$$

where E_F is the Fermi level energy, k_0 is the Boltzmann constant, and T is the absolute temperature.

Consider a crystal with the bounded coordinate $z \in [0, L^*]$ and the infinite coordinates $x, y \in]-\infty, +\infty[$. In the frame of the one-particle Hartree approximation the wave function may be presented as

$$e^{i\mathbf{k}_\parallel \mathbf{r}_\parallel} \psi_i(z, k_\parallel),$$

where $\psi_i(z, k_\parallel)$ is the enveloping wave function for the bounded state $E_i(k_\parallel)$. Here $k_\parallel^2 = k_x^2 + k_y^2$, k_x, $k_y \in]-\infty, +\infty[$; $r_\parallel^2 = x^2 + y^2$. After the change of variables $\varepsilon_\parallel = \hbar^2 k_\parallel^2/(2m_e k_0 T)$, $\varepsilon_i = E_i/(k_0 T)$, $\varepsilon_F = E_F/(k_0 T)$, and $\varepsilon_g = E_g/(k_0 T)$ the concentrations of the electrons and heavy holes can be found from the expressions

$$\rho_e(z) = \frac{m_e k_0 T}{\pi \hbar^2} \int_0^\infty \sum_i \frac{|\psi_i(z, \varepsilon_\parallel)|^2}{1 + \exp(\varepsilon_i - \varepsilon_F)} \, d\varepsilon_\parallel, \qquad (7.3)$$

$$\rho_{hh}(z) = \sum_{j=1}^{+\infty} \Gamma_j(\varepsilon_j) |\psi_j(z)|^2, \qquad (7.4)$$

where

$$\Gamma_j(\varepsilon_j) = \frac{m_{hh} k_0 T}{\pi \hbar^2} \ln[1 + \exp(-\varepsilon_j - \varepsilon_g - \varepsilon_F)].$$

The electrostatic potential $V(z)$ in the SCL created by an external electric field satisfies the Poisson equation [190]

$$\frac{d^2 V(z)}{dz^2} = q \cdot \frac{\rho_e(z) - \rho_{hh}(z) + N_a - N_d}{\varepsilon_0 \varepsilon_{sc}}, \tag{7.5}$$

$$V(0) = V_s, \qquad V(L^*) = V^0(L^*).$$

Here ε_{sc} is the static dielectric constant of the semiconductor; N_d and N_a are the concentrations of ionized donor and acceptor impurities, respectively.

In terms of the quantum description in the one-particle Hartree approximation the wave functions $\psi_i(z, k_\parallel)$, $\psi_j(z)$ and eigenvalues $E_i(k_\parallel, E_j)$ of the free charge carriers are derived from solving the equations

$$-\frac{\partial^2 \psi(z, k_\parallel)}{\partial z^2} = \{k^2 [E - V(z)] - k_\parallel^2\} \psi(z, k_\parallel), \tag{7.6}$$

$$\left(-\frac{\hbar^2}{2m_{hh}} \frac{d^2}{dz^2} - qV(z)\right) \psi_j(z) = E_j \psi_j(z). \tag{7.7}$$

The boundary conditions and normalization of all wave functions take the form

$$\psi(0) = \psi(L^*) = 0, \qquad \int_0^{L^*} |\psi(z)|^2 \, dz = 1.$$

The densities of charge carriers are calculated according to Section 8.3. In our case at room temperature the integration is performed in the interval $\varepsilon_\parallel \in [0, 16]$ taking into account all quantum subbands up to $|\varepsilon_i(k_\parallel = 0)| < 12$ and $|-\varepsilon_j - \varepsilon_g| < 12$ for electrons and holes, respectively. The self-consistent calculation of SCL characteristics is carried out in the frame of the iteration scheme [190].

The following constants of the material [116] were used for calculations: $\varepsilon_{sc} = 17.9$; $m_0 = 9.1 \cdot 10^{-31}$ kg; $m_e = 0.013 m_0$; $m_1 = 0.5 m_0$; $m_2 = 0.015 m_0$; $N_{a,d} = 0$; $E_g = (0.165{-}2.8) \cdot 10^{-4} (T - 300)$ eV; $T = 290$ K. The capacitance-voltage dependence is an important measurable characteristic. The differential capacitance of the SCL is given by

$$C_{sc}(V_s) = \frac{dQ}{dV_s}, \qquad \text{where } Q = q \cdot \int_0^\infty [\rho_e(z) - \rho_h(z) + N_a - N_d] \, dz.$$

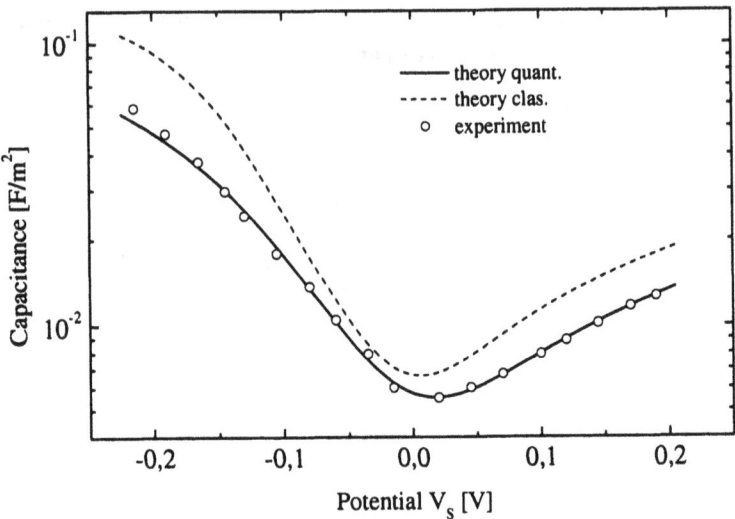

Figure 7.1 Capacitance-voltage characteristics.

The comparison between experimental data, theoretical quantum calculation, and classical calculation for InSb at room temperature is shown in Figure 7.1 [191].

The good agreement of the experimental and quantum results in a wide range of surface potentials demonstrates the efficiency of the proposed mathematical model.

This model of calculation allows one to carry out a detailed analysis of the change of the electron mass in the SCL for the Kane semiconductor [191]. The mass of the electron in the longitudal direction is described by the formula

$$\frac{1}{m_i(E_\parallel)} = \frac{1}{\hbar^2 k_\parallel} \frac{dE_i}{dk_\parallel} = \frac{1}{m_e} \frac{dE_i}{dE_\parallel}.$$

We determine the mean value of the electron mass $\overline{m}(z)$ as

$$\overline{m}(z) = \frac{\int\limits_0^\infty d\varepsilon_\parallel \sum_i \dfrac{m_i(\varepsilon_\parallel)|\psi_i(z, \varepsilon_\parallel)|^2}{1+\exp(\varepsilon_i - \varepsilon_F)}}{\rho_e(z)}.$$

One can see from Figure 7.2a [191] that near the surface the concentration of "heavy" electrons is greater than in the bulk. The average

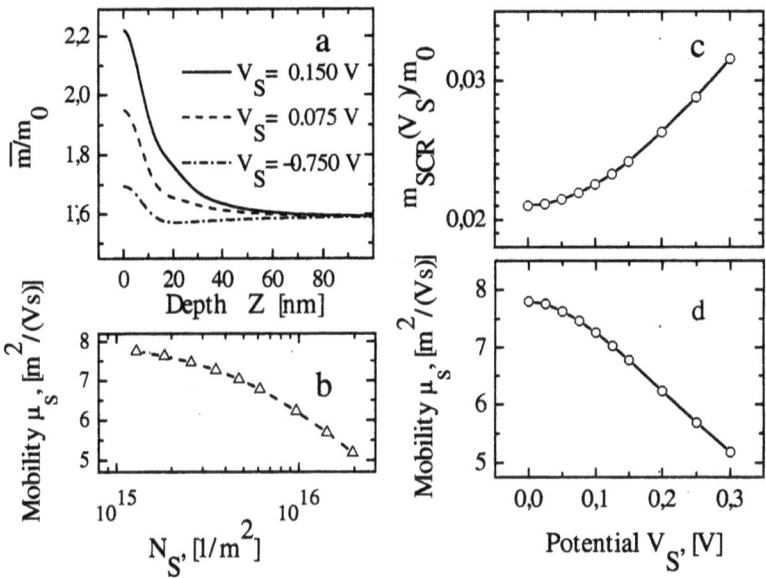

Figure 7.2 (a) Average mass ($\overline{m}$) of the "heavy" electron near the surface; (b) mean electron mobility of electron in confinement in dependence on their concentration; (c) mean mass of the electron in the SCL; and (d) mean electron mobility in dependence on the surface potential.

mass of electron of the i-energy level is described by the formula [191]:

$$\overline{m}_i = \frac{\int\limits_0^\infty \dfrac{m_i(\varepsilon_\parallel)\, d\varepsilon_\parallel}{1+\exp(\varepsilon_i - \varepsilon_F)}}{\int\limits_0^\infty \dfrac{d\varepsilon_\parallel}{1+\exp(\varepsilon_i - \varepsilon_F)}}.$$

Let us introduce the mean mass of the electron in the SCL as

$$m_{\mathrm{SCL}}(V_s) = \frac{\int\limits_0^{2L_D} \overline{m}(z)\rho_e(z)\, dz}{\int\limits_0^{2L_D} \rho_e(z)\, dz}.$$

The correspondent electron mobility is shown in Figure 7.2b, d.

It is obvious that the mean mass of the electron increases monotonically with growth of the surface potential. This regularity must

113

arise also in transport phenomena in subsurface layers of Kane semiconductors.

7.2 TWO-DIMENSIONAL SIZE QUANTIZATION IN SEMICONDUCTOR-ELECTROLYTE INTERFACES AND ITS MANIFESTATION IN FESE EXPERIMENTS

Size quantization in the SCL region at room temperature may be revealed in capacitance-voltage and current-voltage curves as well as in the dependence of surface conductance on the electrode potential. It is manifested in the capacitance (conductance) decrease as compared with the classical case and also in the occurrence of a steplike structure in the curves [179]. Typically, several quantum subbands are filled in narrow-gap semiconductors and also the continuous spectrum and thermal broadening play a significant role. As a result, the steplike structure may essentially be flattened and is revealed as a broadening of the experimental capacitance-voltage characteristics and the dependence of the surface conductance on the electrode potential.

A size quantization effect in a SE system was evidently first observed in ZnO [193] and Si [194]. The existence of two-dimensional quantum subbands for electrons in CdTe at the electrolyte boundary was substantiated by photoluminescence and electroreflectance evidence [195]. The wavelength dependence of the electroreflectance coefficient shows strong oscillations in that part of the spectrum where a maximum photoluminescence is attained (Fig. 7.3). The spectral region is narrow and shifted toward shorter wavelengths with increasing surface potential. These findings, together with the characteristic features of oscillations and the temperature interval for their observation, point to the fact that they are due to surface excitons, when an electron belongs to the two-dimensional surface subband.

Analogous results were obtained in studies of the electroreflectance spectra in the InSb–aqueous electrolyte system (see Fig. 7.4). The energy position of most of the electroreflectance peaks is in good agreement with calculations of the size-quantized subbands for InSb thin films of different thicknesses [196].

A nonmonotonic dependence of the photoemission current on the potential drop at a Bi film–0.5 mol KCl solution boundary due to the

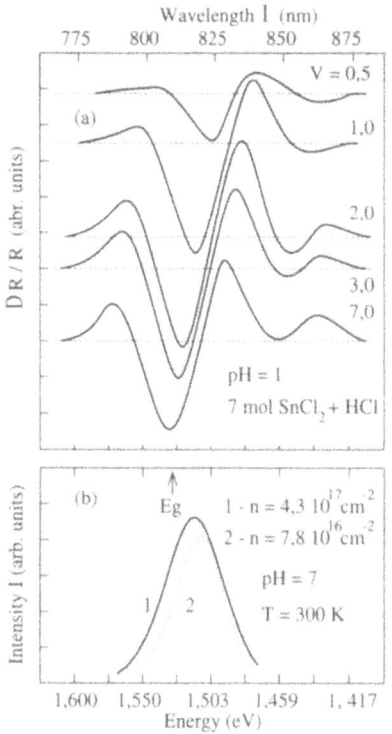

Figure 7.3 Wavelength dependences of the electroreflectance coefficient (a) for CdTe-electrolyte boundary (b) and of the photoluminescence.

size quantization of charge carriers in Bi was found in [197, 198] (Fig. 7.5).

According to the authors of that work, the steps observed in the current-voltage dependences are the result of quantization of charge carriers in the direction normal to the film surface.

The occurrence of a fine steplike structure in the electronic branch of the capacitance-voltage characteristics observed in GaSb electrodes (see Fig. 5.13, Section 5.2.1) may also be considered as a manifestation of size quantization [199]. The positions of the steps in the curves experimentally obtained from the potential magnitudes, calculated from the potential magnitude corresponding to the flat-band potential, agree well with the positions of the first two quantum subbands, calculated within the triangle barrier potential approximation [179]. The experiment yields electrode potential values corresponding to the ground (ψ_0) and the first excited (ψ_1) quantum subbands, equal, respectively, to 0.48 and 0.70 V; the calculated values are 0.40 and 0.75 V, respectively. The two-dimensional electron state

115

Figure 7.4 Top: Dependence of electroreflectance spectrum of the system InSb film–1% NaCl; film thickness 100 Å. Bottom: Dependence of optical absorption spectrum for InSb films on their thickness h (Å): (1), 100; (2), 70; (3), 50. Arrows show subband absorption.

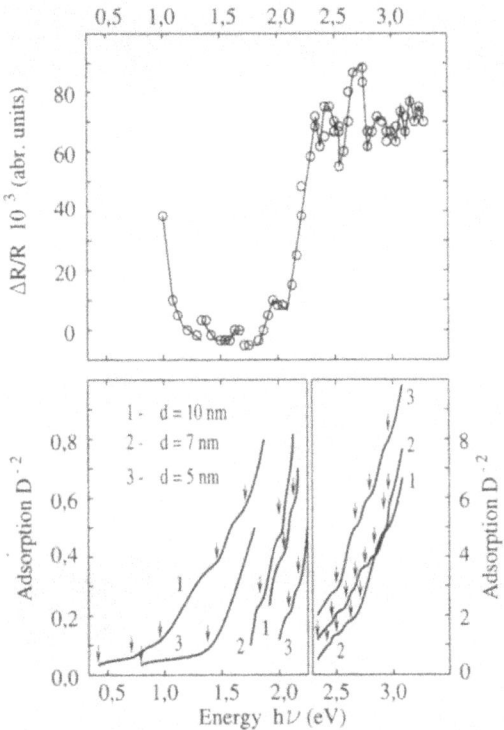

Figure 7.5 Dependence of the photoemission current on the potential drop at the Bi film–0.5 mol KCl solution boundary for different thicknesses of the film.

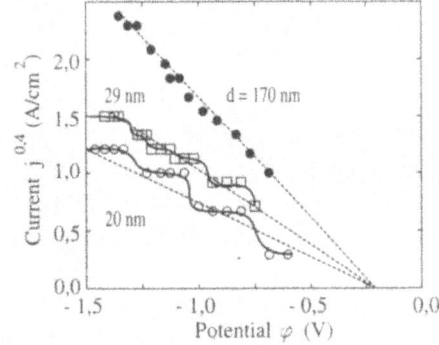

densities in the ground and the first quantum subbands defined from experiment and the calculated ones are also close in value ($5.85 \cdot 10^{11}$ cm^{-2} and $5.25 \cdot 10^{11}$ cm^{-2}, respectively).

The authors of [142, 143] claim that the occurrence of a fine steplike structure in the experimental $C(\varphi)$ and $dN/dE(\varphi)$ curves in the HgTe-electrolyte system (Figs. 7.6 and 7.7) is due to the size quantization effect in the zero-gap semiconductor SCL.

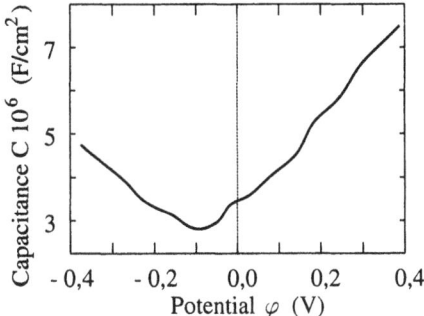

Figure 7.6 Capacitance-voltage characteristic of the HgTe electrode in KCl electrolyte.

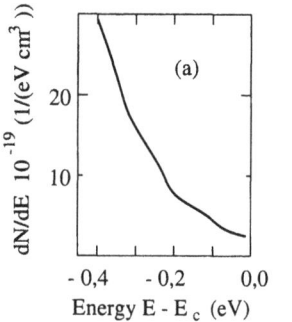 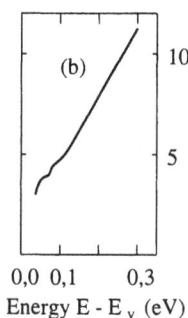

Figure 7.7 Differential density of electron states in c (a) and v bands (b) as functions of the electron energy for HgTe electrode.

The feasibility of observing this effect at room temperatures is due to the low values of the electron effective mass, to the high initial concentration of free carriers in these compounds, and also to the high electric field at the semiconductor-electrolyte interface. This provides a localization of electrons in the near-surface region whose thickness is comparable to the electron wavelength. It is of prime importance that a steplike structure for both $C(\varphi)$ and for $C(T)$ curves in the case of HgTe is characteristic not only for the electron but also for the hole branch of the capacitance-voltage characteristic (Figs. 7.6 and 5.38b). This allows us to suggest that, with surface potentials corresponding to hole accumulation in the SCL in a zero-gap semiconductor-electrolyte system, an appreciable input is due to the light hole band. This conclusion should be considered as additional evidence in favor of the fact that light holes are essential in the characterization of kinetic effects in zero-gap semiconductors [88].

Low-dimensional effects in HgTe are revealed also in the dependence of the differential density of electron states on energy (Fig. 7.7).

A manifestation of quantum size effects in the graphite fiber–electrolyte system was demonstrated in [152]. The diameter of the

Figure 7.8 Differential capacitance (a) and surface conductivity (b) as functions of the electrode potential for the system graphite fiber – 2 mol KCl aqueous solution.

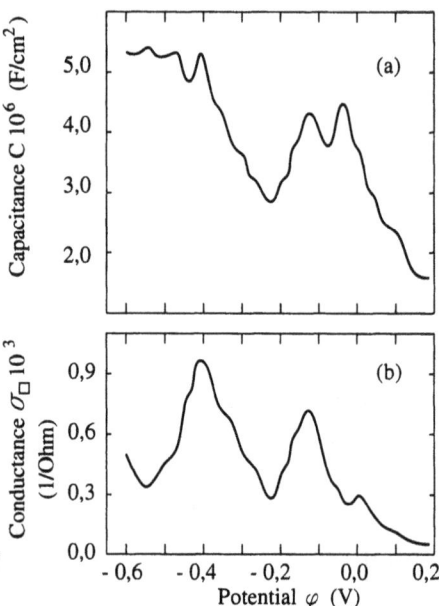

fiber was $10\,\mu$m; $\sigma(\varphi)$ and $C(\varphi)$ curves for such a system are illustrated in Fig. 7.8. The fine steplike structure in the capacitance-voltage curve can be explained [152] by taking into account the size quantization (two-dimensionality) effect of the electron gas in the graphite SCL. In order to evaluate the practicability of this effect, the authors of [152] suggest that the capacitance and conductance drops are associated with the filling of the next band. In this case the carrier density per unit square increases by the value of

$$\Delta n_s = m_e \frac{\Delta E}{\pi \hbar^2} \simeq 5.5 \cdot 10^{15}\,\text{m}^{-2}, \qquad (7.8)$$

where $m_e \simeq 10^{-2} m_0$, and $\Delta E = q\,\Delta\varphi = q\,\Delta V_s$ is the energy increment corresponding to the first step.

Using the measured value of the conductance, one can make an estimate of the electronic mobility making use of the formula $\mu_n \simeq \Delta\sigma_\square/(q\,\Delta n_s) \simeq 8 \cdot 10^3\,\text{cm}^2/\text{V}\cdot\text{s}$, and of the mean value of the differential capacitance in the interval ΔV_s, by the following formula: $\overline{C} \simeq \Delta\sigma_\square/(\mu_n \Delta V_s)$, where $\Delta\sigma_\square$ is the conductance change in this interval. The estimate for the first step yields $C \simeq 2.8 \cdot 10^{-2}\,\text{F/m}^2$, which is in good agreement with experiment (see Fig. 7.8a).

The authors of [200, 201] suggested an original way of investigating the electronic properties of two-dimensional systems on a semiconductor surface making use of the SE boundary. A drop of an electolyte was placed onto the Ge plate and held in place by the mica platelet. As a result of the electrolyte creep, a capillary layer of 0.0001 cm thickness was produced between the surfaces. The mica platelet was held by means of the capillary pressure force. An electrochemical cell with a 14 mol HNO_3 solution was cooled down to liquid helium temperature. In so doing, the double electric layer at the SE interface was frozen. The relevant charge localized in the double layer produced an electric field in the SCL sufficient to form a quantum potential well for equilibrium carriers in the near-surface region of the semiconductor. It was found that in this region, spatially separated two-dimensional layers of electrons and holes arise. Herewith, the surface of germanium with a great number of surface states features a phase transition resulting in the formation of a two-dimensional electron-hole condensate, revealed in the photoluminescence spectra as a novel line due to the recombination radiation.

7.3 ONE-DIMENSIONAL QUANTUM SYSTEM: THEORY

We shall consider a quantum wire (QW) with sizes in the coordinates x, y about the de Broglie wavelength of an electron, and with great extent along the axis z [182, 189]. Then the quantity of energy of an electron relevant to motion along the z axis varies quasicontinuously, and is quantized along the x and y axes (see Fig. 7.9).

The bulk dispersion law of Kane semiconductors is described by formula (8.4). In the frame of the one-particle Hartree approximation the wave function may be presented as $e^{ik_z z}\psi_i(x, y, k_z)$ where $\psi_i(x, y, k_z)$ is the envelope wave function for the bounded state $E_i(k_z)$, $x \in [0, L_x]$, $y \in [0, L_x]$, and $z \in]-\infty, +\infty[$.

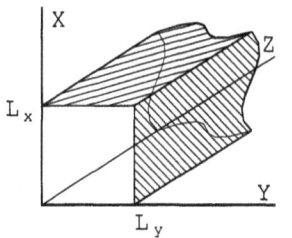

Figure 7.9 Approximate model of the quantum wire.

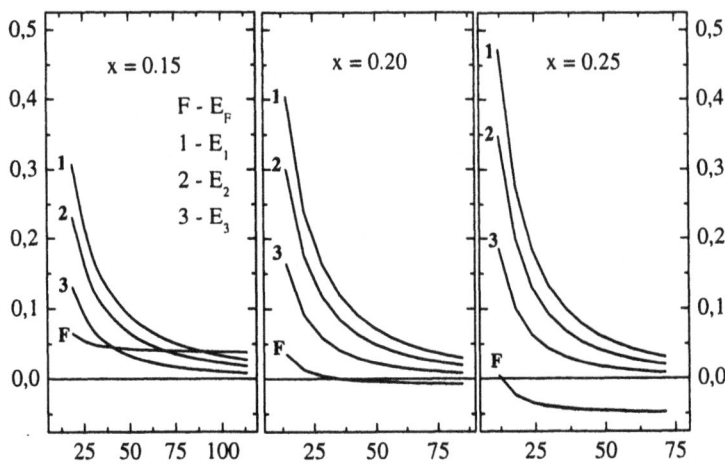

Figure 7.10 Distance between energy levels for several CMT compositions $(x = 0.15, 0.20, 0.25; T = 295\,\mathrm{K})$.

In the general case, to find the standing discrete quantum states of energy $E_i(k_z)$ it is possible to use the expression for complete quasi-momentum

$$\frac{\partial^2 \psi_i(x, y, k_z)}{\partial x^2} + \frac{\partial^2 \psi_i(x, y, k_z)}{\partial y^2} = \{k^2[E_i(k_z)] - qV(x, y) - k_z^2\}\psi_i(x, y, k_z).$$

(7.9)

In the case of absence of an exterior electric field $[V(x, y) = 0]$ this equation is reduced to the well investigated equation

$$\frac{\partial^2 \psi_i(x, y)}{\partial x^2} + \frac{\partial^2 \psi_i(x, y)}{\partial y^2} = -\lambda \psi_i(x, y).$$

(7.10)

The maximum distance between levels is gained in the case $L_x = L_y$. The results of calculations for various compositions of CdHgTe and two temperatures are shown in Fig. 7.10.

The lines correspond to the bottom of the quantum subbands ($k_z = 0$). The ratio of the quantity of carriers in the first quantum subband to the total of electrons and also the electron concentration in the quantum subbands are shown in Figure 7.11.

From the analysis shown in Figures 7.11 and 7.10 it is possible to make some recommendations about the choice of geometrical sizes for manufacturing quantum wires in the shape of rectangular "ingots"

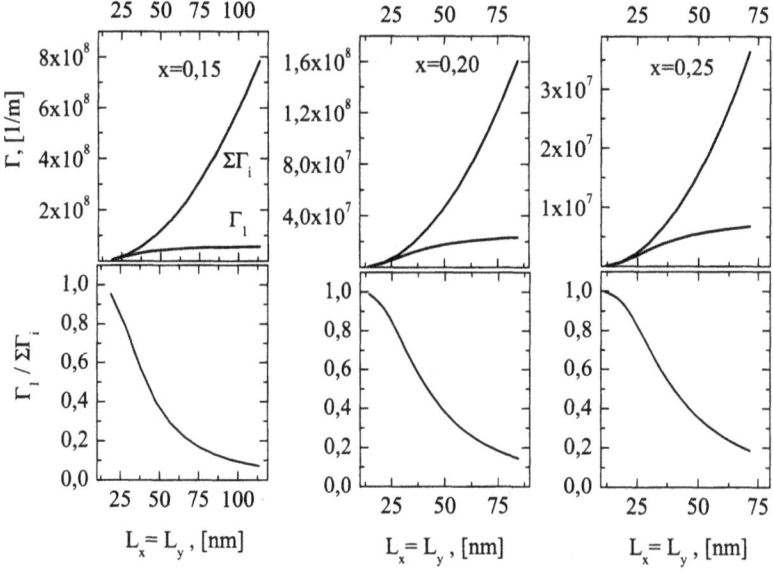

Figure 7.11 Dependences of electron concentration in quantum subbands for several CMT compositions ($x = 0.15, 0.20, 0.25$; $T = 295$ K).

from CdHgTe for its three compositions. Assuming that the length of a wire is comparable to the free length of an electron in CdHgTe, it is possible to estimate the electron concentration in a quantum wire. It is necessary to note that the given calculation is only an estimate and can be considered only as a first approach. Certainly a more precise calculation must be based on the analysis of quantum wires taking into consideration their geometry and shape obtained in experiments.

7.4 FABRICATION OF LOW-DIMENSIONAL QUANTUM STRUCTURES AND THEIR ELECTRONIC PROPERTIES

This time systems with low-dimensional electron gases are formed mainly in heterostructures, which are synthesized on the basis of SiGe-Si and (AlGa)As-GaAs combinations by way of molecular-beam epitaxy [181]. Electron gas dimension lowering occurs because of its location in the space charge layer of a rough heterojunction. The obstacles to observing the quantum properties of an electron gas are the effects of heat and collisional widening of the discrete energy levels, which generally occur at room temperature. In connection with this, one physical task is the development of methods allowing structures, to be obtained whose properties would be determined by quantum dimensional effects up to room temperature. It is possible to point out two basic directions of investigations allowing solution of this problem.

1. Use of materials with small effective mass and high mobility of electrons and correspondingly with large wavelength and free path.

2. Creation of new methods of forming low-dimensional structures by the possibility of simultaneous control of their electronic properties, that is, in situ.

Quantum low-dimensional structures may be successfully created on the basis of materials that are solid solutions of cadmium, mercury, and tellurium (CMT solutions). The basic parameters of these materials for four different compositions are shown in Tables 7.1 and 7.2. The important peculiarity of CMT combinations

TABLE 7.1

Electronic Characteristics of $Cd_xHg_{1-x}Te$

x	a_0 (nm)	m_e/m_0	E_g (eV)	m_h/m_0	ε_{sc}	μ_e $(cm^2(V \cdot s)^{-1})$
0.15	0.64633	0.007	0.080	0.50	18	32,000
0.20	0.64641	0.020	0.150	0.45	17	20,000
0.32	0.64663	0.030	0.320	0.45	16	7,000
References	[131, 180]	[131, 180]	[131, 180]	[131, 180]	[180]	[131, 202]

Here: E_g is the band gap; m_h the effective mass of holes; m_0 the mass of the free electron; and ε_{sc} the dielectric constant of semiconductor.

TABLE 7.2

Quantum Parameters of the Electron in the CMT Material at Room Temperature

$Cd_xHg_{1-x}Te$	$\langle m_i \rangle / m_0$ $(i = 1)$	L_0 (nm)	λ_e (nm)
$x = 0.15$	0.007	300	59
$x = 0.20$	0.012	200	55
$x = 0.32$	0.023	80	45

is the great dependency of their characteristics on the composition. That allows creation of heterostructures by way of formation of contacts between the regions of CMT material with different compositions. Thus a small change of the lattice constant a_0 on changing the CMT composition (see Table 7.1), and consequently the weak destruction of the atomic structure period in the heterojunction region, leads to a low density of surface states and as a consequence, to a small value of electron capture and scattering.

The change of the CMT material composition and simultaneous measurement of electrophysical properties of formed structures may be performed in the electrochemical cell by help of the field effect in an electrolyte. The basis of this effect is the change of the electrode (V) and accordingly of the surface (V_s) potentials of a semiconductor electrode by its polarization in an electrolyte [16]. As applied to CMT materials the possibility of this manner of material composition change was shown in [182, 203].

7.4.1   Experiment

As the initial CMT material we used samples with compositions corresponding to $x = 0.32$ (see Table 7.1). First the surface of the samples was exposed to chemical dynamic etching in 8% bromine solution in methanol [127]. To change the composition of the CMT material a complexing agent was introduced into the measuring cell containing the aqueous electrolyte. The complexing agent was selected so that the cadmium complexes appearing were stable in a larger potential interval than the mercury complexes. As a result in some interval of electrode potentials a deficiency of cadmium appeared near the sample surface because of dissolution of cadmium complexes. As a consequence, a layer formed near the surface of the sample enriched by mercury. The composition of this layer corresponded to $x < 0.32$ and it had a smaller energy gap than the initial material. As a result a two-layer heterostructure formed, which had a layer of narrow-gap semiconductor on the surface of the initial CMT material. The thickness and the composition of this layer were determined with the help of capacitance-voltage characteristics [$C(\varphi)$ characteristics] measured by methods suggested in [204] (see also Section 8.1). Additional control of the surface layer of the narrow-gap semiconductor was carried out with the help of IR absorption. The capacitance-voltage characteristics of the structure were measured directly in the

electrochemical cell using the four-electrode scheme [17] by potentiostatic set of the electrode potential. To measure we applied an
impulse signal 0.1 ms long and used the middle value of not less than
32 pulses for each potential value. Moreover for each testing impulse
we measured the time constant of the potential relaxation [$t(\varphi)$] for
the electrode-electrolyte system. This quantity and measured capacitance [$C_{int}(\varphi)$] was used for estimation of the value of the shunting
resistivity of the structure [$R_{int}(\varphi) = t(\varphi)/C_{int}(\varphi)$]. The IR absorption was measured on a dry structure at room temperature in a dry
nitrogen atmosphere. The reverse optical scheme with a 10–15 Hz
monochromatic beam modulation was used. The signal registration
was performed by using a differential bolometer with the help of a
variable current bridge having a time of accumulation equal to 32 s for
each wavelength value.

7.4.2 Heterostructures Forming Two-Dimensional Electron Gases and Their Electrophysical Properties

The capacitance-voltage characteristics of the surface of a CMT electrode for two different samples obtained by the potentiostatic change
of the electrode potential are given in Figure 7.12. This figure contains
also the calculated dependence of the space charge layer capacitance
on the surface potential for the CMT material corresponding to a
composition with $x = 0.32$. There are the following intervals which
correspond to different stages of structure formation. The interval
of electrode potential from +50 to −250 mV corresponds to dissolution of the electrode surface with conservation of the initial correlation between all components of the triple combination (interval "A" in
Fig. 7.12). It is marked by the reversibility of the $C(\varphi)$ characteristics
and their conformity to capacitance change in the space charge layer
of the CMT material with initial composition $x = 0.32$. In the interval
of electrode potential less than −250 mV mercury complexes begin to
discharge on the CMT electrode surface while cadmium complexes
stay stable and continue freely to go out into solution. Thus the electrode surface is enriched with mercury. This process is followed by
the irreversible decrease of capacitance with subsequent stabilization of $C(\varphi)$ characteristics in the electrode potential region from
−350 to −550 mV (region "B" in Fig. 7.12). In the region from −600 to
−750 mV the reversible change of $C(\varphi)$ characteristics appears again
(region "C" in Fig. 7.12). For potentials less than −800 mV irreversible

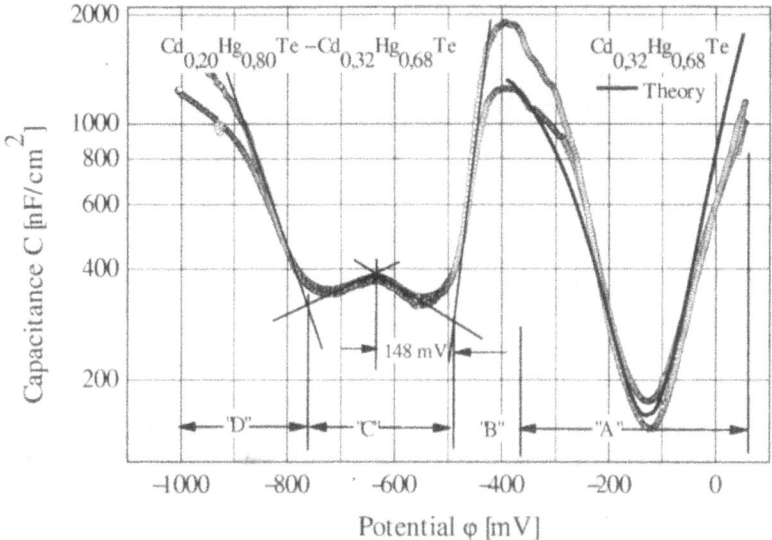

Figure 7.12 Two typical capacitance-voltage characteristics of the modification process for $Cd_{0.33}Hg_{0.67}Te$ substrate.

destruction of the electrode materials takes place. The decrease of the capacitance when the potential is in interval "C" and the reversible character of the $C(\varphi)$ characteristic change in this interval may be interpreted as the appearance of a new stable structure, contained inside the insulator layer. An evaluation of the thickness of this layer from the experimental value of the capacitance is 36–41 nm, indicating that the dielectric constant of this layer (ε_{sc}) is 16–18, which is characteristic for CMT materials (see Table 7.1). This layer thickness is comparable with the electron wavelength in CMT materials with composition corresponding to $x = 0.20$ (see Table 7.1), which may be the cause of the appearance of electron gas dimensional quantization in the region of the considered heterostructure.

The results of IR absorption measurement for the formed structure are given in Figure 7.13. There are two peculiarities in the presented spectrum at wavelengths of 3900 and 7800 nm. These peculiarities testify to the formation of a two-layer structure, which consists of an underlayer with composition corresponding to $x = 0.32$ and a semitransparent layer with composition corresponding to $x = 0.20$. The thickness of this layer evaluated on the basis of IR absorption taking into consideration the refraction coefficient $n = 3.6$ [131] is

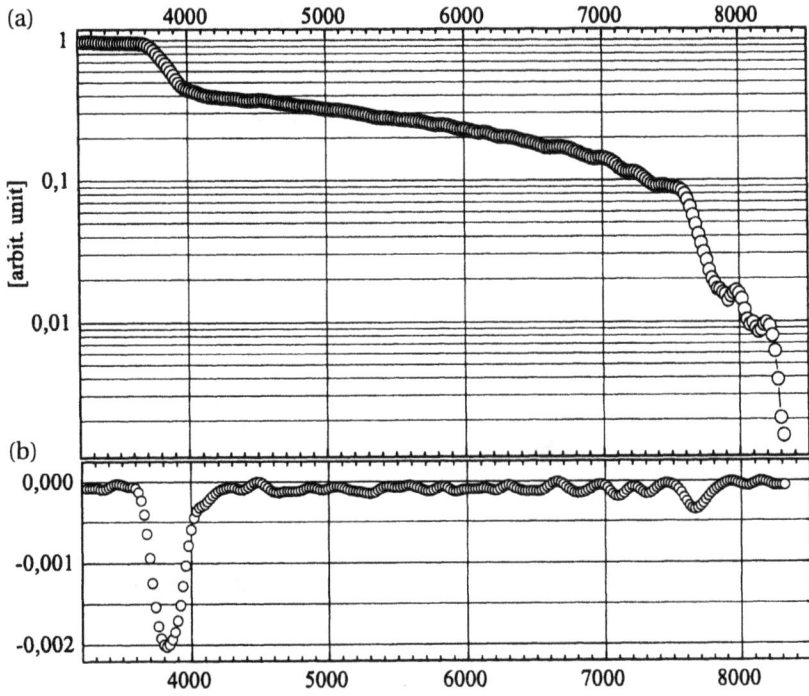

Figure 7.13 Absorption spectrum (a) and its derivation (b) for a sample with a modified layer on $Cd_{0.33}Hg_{0.67}Te$ substrate.

from 200 to 400 nm. This evaluation gives a thickness of the layer five times greater than the thickness of the insulator layer calculated from the $C(\varphi)$ characteristics. The results of differentiation of the absorption spectrum curve are given in Figure 7.13. The fine structure of the absorption spectrum edge corresponds with the existence of a quantum size electron gas in the region of the semitransparent layer. The results obtained indicate the formation of a heterojunction, appearing in the contact of the initial underlayer with a CMT layer having the composition corresponding to $x = 0.20$.

The process of the formation of the heterostructure may be explained in Figure 7.14. The surface potential of the initial underlayer is defined by the electrochemical reaction of $Cd_{0.32}Hg_{0.68}Te$ disso-lution. The change of potential in the polarization process will be limited by the potential of mercury complex discharge, followed by the appearance of atomic mercury (Hg^0) on the electrode surface (see Fig. 7.14a). The appearing mercury will lightly penetrate into

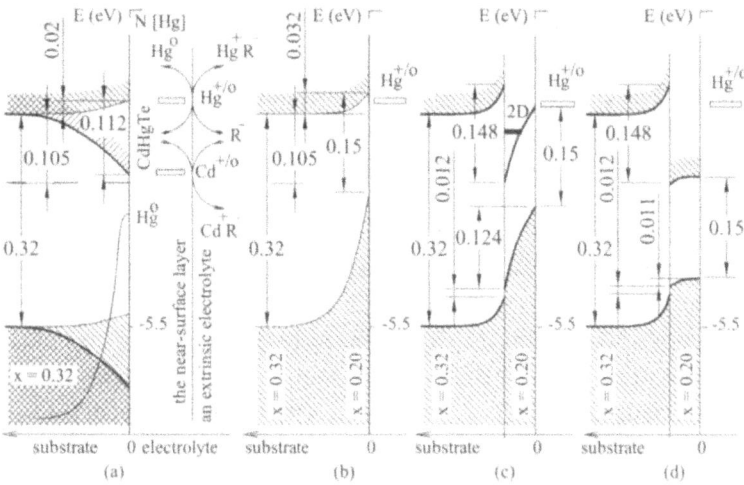

Figure 7.14 Band diagram for a sample with a modified layer on $Cd_{0.33}Hg_{0.67}Te$ substrate.

the CMT bulk near the surface, changing its composition. A similar process of CMT electrode composition change was earlier described in [203]. The change of CMT electrode composition near the surface leads to change of its corrosion potential, which involves a change of its surface potential. So in the electrolyte-$Cd_{0.20}Hg_{0.80}Te$-$Cd_{0.32}Hg_{0.68}Te$ system the potential in the $Cd_{0.32}Hg_{0.68}Te$ electrode volume and the surface potential of $Cd_{0.20}Hg_{0.80}Te$ become fixed. The energy band structure of the system surface layer of composition $Cd_{0.20}Hg_{0.80}Te$-underlayer with composition $Cd_{0.32}Hg_{0.68}Te$ may correspond to a graded-band structure (see Fig. 7.14b) or rough heterostructure (see Figs. 7.14c and 7.14d). However, graded-band character of the structure is in contradiction to the IR absorption experiments (Fig. 7.13) and to the appearance of the insulator layer, which is shown by the $C(\varphi)$ characteristics (Fig. 7.12). Simultaneously, the model of the rough heterojunction allows one to explain the appearance of oscillations near the absorption edge of the layer (Fig. 7.14b) and the decrease of the capacitance in the region "C" because of the electron gas size quantization, which takes place in the region of breaking energy bands in the heterojunction.

Returning to Figure 7.12 one notices that the right branch of the $C(\varphi)$ characteristics extrapolated by straight lines corresponds to the formation of a $Cd_{0.20}Hg_{0.80}Te$ surface layer, and the left one characterizes the change of full capacitance of the heterojunction by its polarization.

The potential $V_1 = -480\,\mathrm{mV}$ corresponding to crossing the extrapolated parts of the $C(\varphi)$ characteristics differs from the potential of the unbiased junction $V_2 = -628\,\mathrm{mV}$ by the value $V = |V_1 - V_2| = 148\,\mathrm{mV}$ (see Fig. 7.12). This value coincides with the value of the band break in the heterostructure (Fig. 7.14c). That allows us to state that the process of formation of the $Cd_{0.20}Hg_{0.80}Te$ surface layer (see Figs. 7.14a and 7.14b) stops with the appearance of band breaking (see Fig. 7.14c). So two independent methods [the methods of IR absorption and of $C(\varphi)$ characteristics] showed that by electrochemical modification of the surface region of a CMT material a heterostructure appears, formed by contact of two regions of the CMT material with different composition. The band break arising in the contact leads to two-dimensional electron gas appearance in the region of the heterostructure.

Here it is important to note that the essential circumstance in forming a low-temperature heterostructure by this method is the participation of the electron subsystem in the process of the modification of the CMT material. The CMT layer with $x = 0.20$ appearing on the surface of the initial material has a lower value of the electron effective mass and, correspondingly, a larger λ_e as compared with the initial material with $x = 0.32$. Thus formation of the layer corresponding to $x = 0.20$ by the electrochemical dissolution reaction automatically stops when its thickness reaches the value λ_e and the electron subsystem in the layer becomes size quantized. Two mechanisms may play the important roles in this process.

1. In the material being formed the inequality $\lambda_e > a_0$ is satisfied. Thus the probability density of electron localization near one individual atom becomes small. As a consequence the electrochemical process stipulated by the electron-atom interaction and leading to formation of material with a new composition (forming the corresponding chemical bonds) slows down.

2. The electron is localized on the size quantization level. As a result an additional decrease of electron density takes place near the surface. Thus the probability of electron exchange becomes even slower.

Apparently both these mechanisms connected with the wave nature of the electron inhibit the electrochemical reaction which allows formation of layers with thickness comparable with λ_e in the considered material.

7.4.3   Formation and Properties of the
One-Dimensional Conductor

For the formation of a one-dimensional (1D) electronic quantum structure (quantum wire) the interface between the CMT material ($x = 0.32$) and electrolyte, simultaneously with polarization, was subjected to the influence of ultrasound vibrations, which caused the appearance of whirlwind streams of liquid in this interface (see Fig. 7.15). The action of whirlwind streams led to the diffusion limitations being locally removed, which promoted in addition the dissolution of cadmium. As a result, simultaneously with modification of the CMT composition as described in Section 7.4.2 the formation of CMT material regions occurred whose composition corresponded to $x < 0.20$. The appearance of these regions as a geometrical heterogeneity in its turn led to the stabilization of the whirlwind streams. Examples of the enforced inhomogeneity are shown in Figure 7.16. As the ultrasound strength increases a network of inhomogeneities appears independent of the external edge. In this way a honeycomb-like structure was formed, presented in Figure 7.17. In our opinion this process has some analogy with formation of Benar-Tailor cells,

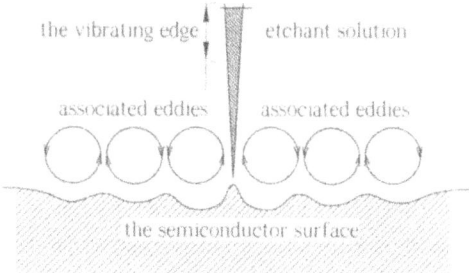

the vibrating edge   etchant solution

associated eddies   associated eddies

the semiconductor surface

Figure 7.15 Profile of the geometrical inhomogeneity emerging in the region of the eddies near the vibration edge.

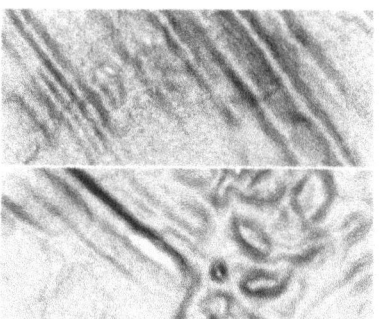

Figure 7.16 Reconstructed optical image of oblong geometrical structures of the modified material, which are evoked in the middle of the vibrating edge and images of the same structures at the end of the vibrating edge. Both scales are approximately 400 nm per 1 mm.

Figure 7.17 Typical images of region containing the honeycomb lattice at the final stage of etching. The scale is approximately 0.025 mm per 1 mm.

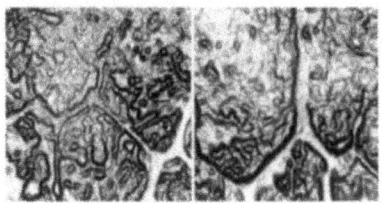

Figure 7.18 Image of the node of a honeycomb lattice. The scale is 25 μm per 1 cm. The geometric inhomogeneity with 1D wires. The scale is 500 nm per 1 mm.

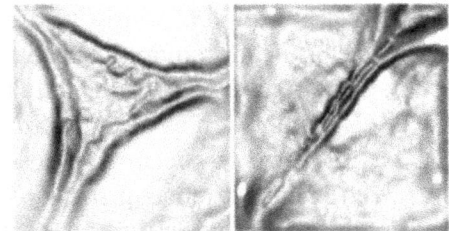

which are typical examples of a self-organization process [45]. One-dimensional quantum conductivity occurred in threadlike fragments of the structure ("whiskers") with cross sections less than 0.005 mm (see Fig. 7.18a) if additional compression of the electronic gas took place by external polarization. The length of the quantum wire created as a fragment of the honeycomblike structure (see Fig. 7.18b) became comparable with the electron free path in $Cd_{0.15}Hg_{0.85}Te$ ($L_0 = 300$ nm at room temperature). This provided ballistic transport of electrons along the wire.

The possibility of one-dimensional conductivity occurring in the honeycomblike structure is based on the fact that the individual elements of the structure presented by CMT materials of different compositions are separated by potential barriers. The height of these barriers was essentially more than the value of kT/q at room temperature (see Fig. 7.19). Figure 7.20a gives a sketch of a fragment of the honeycomblike structure with a 1D wire and its equivalent electric scheme is shown in Figure 7.20b.

The quantum wire in this scheme is represented by the impedance, including a resistor (R_{1D}) shunted by a spread capacitance (C_{1D}) (it is not shown in the figure; see below). The auxiliary platinum electrode in the electrolyte volume plays the role of the entrance contact (3D source) and the volume of the initial CMT sample that of the exit contact (3D substrate). So electrical charge, appearing in knots of the honeycomblike structure, flowed down through the successively

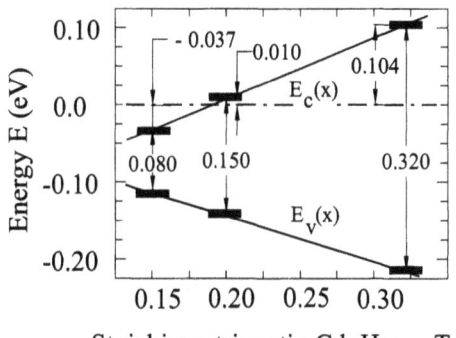

Figure 7.19 Band diagrams for $Cd_xHg_{(1-x)}$Te materials with E_F as the zero level.

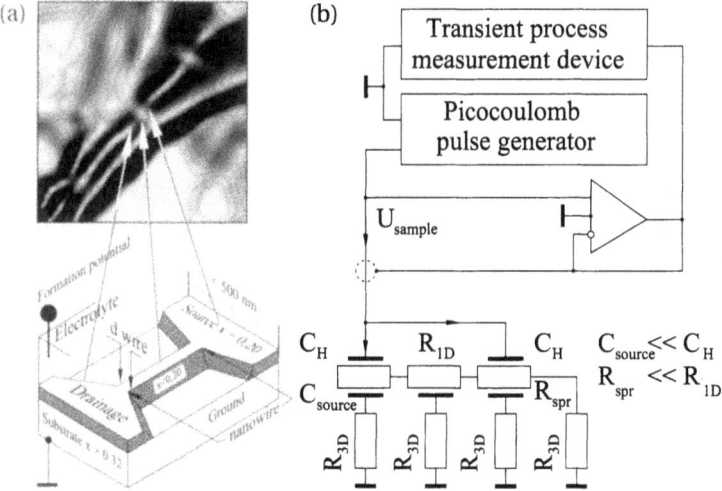

Figure 7.20 (a) Geometric inhomogeneity on the surface of the substrate in the electrolyte. (b) Electric equivalent circuit of the measurement.

connected impedance of the quantum wire and the "spreading" impedance (Z_{spr}). Z_{spr} is the combination of resistors of the regions $Cd_{0.20}Hg_{0.80}$Te and $Cd_{0.32}Hg_{0.68}$Te ($R_{spr} = R_{2D} + R_{3D} \simeq R_{2D}$) and of the capacitance of the heterostructure $C_{source} = C_{2D} \cdot S_{knot} = 200 - 10$ pF, where S_{knot} is the square of a knot of the honeycomblike structure and $C_{2D} = \sum C_{quant}$. The capacitance C_H is that of the electrolyte contact (Helmholtz layer capacitance), which is $C_H \approx 20 \cdot 10^{-6}$ F/cm^2 at $T = 295$ K [3]. The equivalent scheme (Fig. 7.20b) shows that one-dimensional conductivity may be determined by measuring the relaxation (spreading) time of the charge entering the system using the formula

$$G = \frac{C_{source}}{\tau} = -C_{source} \cdot \frac{\ln[U(t)/U_0]}{t_i},$$

where t_i is the duration of the testing impulse of the measuring equipment. Thus $\ln[U(t)/U_0] < 0$, where $U(t)$ and U_0 are the parameters of the structural response tested by the probe impulse. The time constant τ for the structure presented in Figure 7.20 was estimated according to the formula

$$\tau = R_{1D} \cdot C_{source} = 2.3 \cdot 10^{-6} \, s. \tag{7.11}$$

This shows that to select processes connected with charge spreading through a one-dimensional structure it is necessary to carry out measurements in time intervals below 10^{-6} s. The electrophysical characteristics of all other regions of the structure stay unchanged because their time constants are much larger than τ. The values of quantum subband spreading resistance were estimated according to the formula

$$R_{spr} = 1/q \cdot \mu_e \cdot N_i \tag{7.12}$$

where the electron mobility $\mu_e = 20,000 \, cm^2/V \cdot s$ was taken as the bulk value for $Cd_{0.20}Hg_{0.80}Te$ [180, 202]; values of N_i were obtained from the self-consistent solution of the Schrödinger and Poisson equations for semiconductors with nonparabolic dispersion in the conduction band with the help of the method suggested in [190, 191]. The values of quantum subband energy and their electron densities are estimated in [205] and given in Table 7.3.

The results of evaluation of R_{spr} and C_{quant} necessary for the equivalent scheme analysis used are given in this table also. From these data one may see that at a surface potential 150 mV coinciding with the potential in an unbiased junction (see Section 7.4.2) all electronic charge is localized in the first quantum level, which crosses the Fermi level at this potential. The value of the quantum wire resistor was determined from the relation [181]

$$R_{1D} = \frac{h/(2 \cdot e^2)}{n} = \frac{12,900}{n} \quad (\Omega), \qquad n = 1, 2, 3, \ldots. \tag{7.13}$$

TABLE 7.3

Values of Energy and Electron Density in Quantum Subbands in
$Cd_{0.20}Hg_{0.80}Te$ Calculated at Room Temperature, and Volume of
Electronic Charge (N_1) in the first $(i = 1)$ Quantum Subband

V_s (mV)	$(E_1 - E_F)/q$ (mV)	$(E_1 - E_F)/(k_0 \cdot T)$	N_1 (cm^{-2})
25	71	+2.7	$0.9 \cdot 10^{10}$
50	66	+2.5	$1.5 \cdot 10^{10}$
100	23	+0.9	$4.5 \cdot 10^{10}$
150	0	0.0	$11.0 \cdot 10^{10}$
200	−28	−1.1	$21.0 \cdot 10^{10}$

From a comparison of R_{1D} and R_{spr} values one may conclude that $R_{spr} \ll R_{1D}$ for all values of n. This estimation allows us to neglect the value of R_{spr}, which is not shown in Figure 7.20b for this reason. The value of the quantum wire capacitance was estimated according to the following approximate formula:

$$C_{1D} \approx \frac{q^2 N_{i=1}}{E_{i=1} - E_F} \tag{7.14}$$

where $N_{i=1}$ is the density of electrons in the one-dimensional quantum subband with energy $E_{i=1}$. For evaluation of $N_{i=1}$ and $E_{i=1}$ the model of a one-dimensional quantum conductor in the form of a rectangular bar for a CMT material with $x = 0.15$ was chosen (see Section 7.3). Using $L_x = L_y = L_0 = 59$ nm and the formula for C_{1D} one may evaluate the quantum wire capacitance. In the limit of one filled quantum subband $C_{1D} = 1.1$ pF/cm, and at three filled quantum sub-bands $C_{1D} = 3.4$ pF/cm. Comparing the capacitance values given in the equivalent scheme with the quantum wire capacitance, one can see that the latter may be neglected. Thus the capacitance C_{1D} is not shown in Figure 7.20b. The dependence of the conductivity on the voltage at the field electrode for two quantum wire samples is given in Figures 7.21 and 7.22.

In these figures the conductivity is presented as the number of units of the one-dimensional conductivity quantum. The $G(\varphi)$ dependence clearly displays two steps for one of the samples and four steps for the other. The appearance of steps in the dependence of the conductivity on the applied voltage is a direct demonstration of one-dimensional

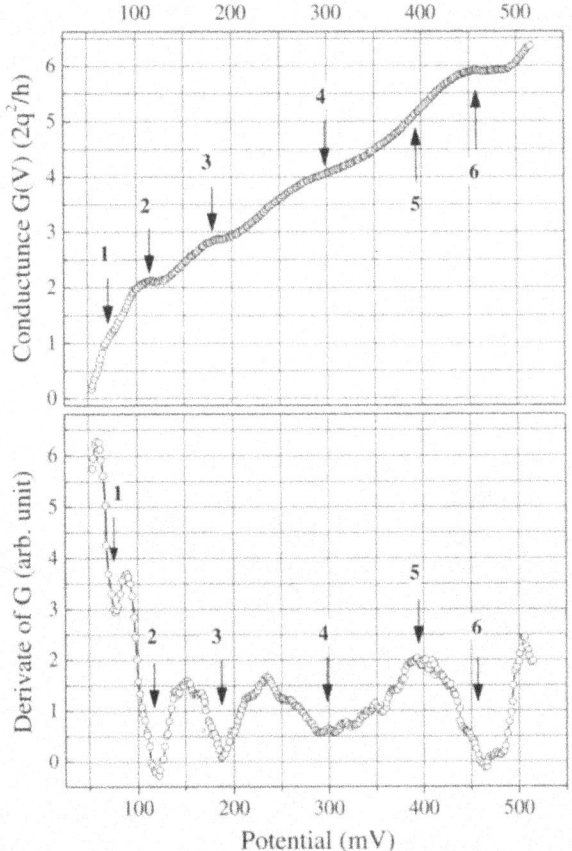

Figure 7.21 G is the conductance normalized to $2q^2/h = 1$. Crosses on the curve mark the experimental data. Derivative of the normalized conductance. Arrows on the curve mark ranges corresponding with integer-valued steps of the normalized conductance.

conductivity and its quantum nature. The remarkable fact is the observation of such conductivity at room temperature. Noninteger values of n are known in the literature [183–188] and need additional study.

7.4.4   Conclusion

As a result of the investigations performed it was shown that structures that display the quantum properties of two-dimensional and one-dimensional electron gases at room temperature may be formed

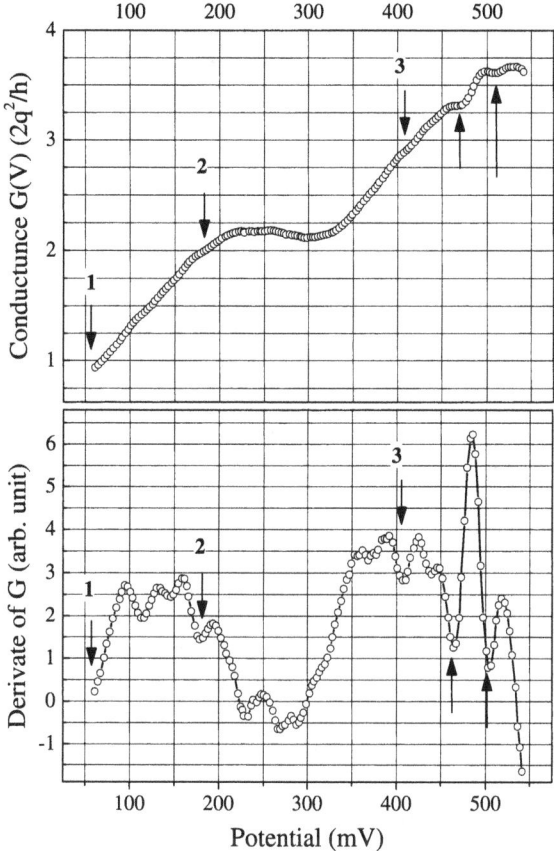

Figure 7.22 G is the conductance normalized to $2q^2/h = 1$. Crosses on the curve mark the experimental data. Derivative of the normalized conductance. Arrows on the curve mark ranges corresponding with integer-valued steps of the normalized conductance.

in the CMT material–electrolyte system by the appropriate choice of the electrolyte and polarization conditions. Self-organization processes may play an essential role in forming these structures. The results showed the possibilities of using semiconductor-electrolyte systems to form low-dimensional quantum structures. This operation includes using the field effect in electrolytes as a forming factor and also as a method for control in situ of the structure forming processes.

Technique of FESE Method and Some Possibilities for Its Technological Applications

8.1 TECHNIQUE OF QUASI-EQUILIBRIUM FESE METHOD

T HE TECHNIQUE for measuring FESE characteristics is described in detail in [130, 206, 207]. In this chapter we shall only point out some peculiarities of the FESE technique which must be taken into consideration in its realization.

The realization of FESE is based on measurements of the impedance of the SE interface, implying extraction of its components (capacitance and conductance), which affect the processes occurring on the semiconductor surface and in the near-surface region. In the range of electrode potentials corresponding to the ideal polarizability of a semiconductor (see Chapter 2), the contribution to the impedance of the SE system is largely defined by the processes occurring in the SCL and in the surface states of the semiconductor. In studies of processes that occur at electrode potentials beyond the ideal polarizability region, the interpretation of results is complicated because of the additional component of impedance caused by electrochemical reactions at the phase boundary [208]. Hence, one of the most important issues in the realization of FESE is proper choice of the electrolyte and the range of electrode potential variations, with regard to the realization of conditions for the ideal polarizability of a semiconductor electrode. Several examples of realization of such conditions are given in Chapters 3 and 6.

Under conditions of ideal polarizability, a physical model of the interface involves two types of charges: Q_{sc}, which corresponds to the contribution from free carriers in the SCL, and Q_{ss}, which is the charge in surface states. They are characterized by relaxation times τ_{sc} and τ_{ss}, respectively. For this case, the equivalent scheme can be represented as in Figure 8.1, where

$$C_{sc} = \frac{\Delta Q_{sc}}{\Delta \varphi}, \qquad R_{sc} = \frac{\tau_{sc}}{C_{sc}},$$

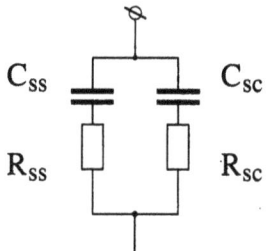

Figure 8.1 Equivalent electric scheme of the semiconductor-electrolyte interface taking into account relaxation times. C_{sc}, R_{sc}, parameters of SCL of the semiconductor; C_{ss}, R_{ss}, parameters of surface states.

$$C_{ss} = \frac{\Delta Q_{ss}}{\Delta \varphi}, \qquad R_{ss} = \frac{\tau_{ss}}{C_{ss}}.$$

In measuring the impedance, the interface impedance is contributed by an additional (supplementary) charge dQ, small in value, which yields the small additional surface potential change dV_s. This does not bring about any essential alteration of the state of the SE system, if the following inequality is satisfied: d$\varphi \ll k_0 T/q$.

Separation of the constituents of impedance measured using FESE that correspond to surface characteristics of the semiconductor can be realized by studying the frequency dispersion of impedance [209–212]. Two approaches can be used, both equivalent from the point of view of the theory of electric circuits. Measurements that make use of a harmonic signal [213] whose frequency does not usually exceed 10^6 Hz are most informative for studies of the impedance frequency dispersion caused by slow ionic processes. These measurements enable monitoring of the electrochemical processes involved in the polarization. The most effective means for defining surface parameters is the analysis of transition processes in the impulse regime [213].

The response of the interface to the current pulse is represented in Figure 8.2, where the dispersion is revealed. The accuracy to which the semiconductor SCL differential capacitance can be defined increases as the probe pulse duration τ_i decreases. Under the condition $\tau_i/\tau_{ss} \ll 1$, the measured capacitance will exactly follow the semiconductor SCL capacitance. Separation of the SCL capacitance and that of surface states by making use of the impedance frequency dispersion is demonstrated in Figure 8.3.

We emphasize that, in measurements of FESE, interaction between different channels of the measuring system should be taken into

Figure 8.2 Response of the semiconductor-electrolyte interface to the current pulse for two probe pulse durations τ_i ($\tau_{i2} > \tau_{i1}$).

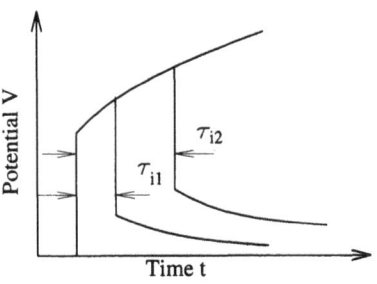

Figure 8.3 Total capacitance of the semiconductor-electrolyte system depending upon pulse duration for the equivalent scheme presented in Fig. 8.1.

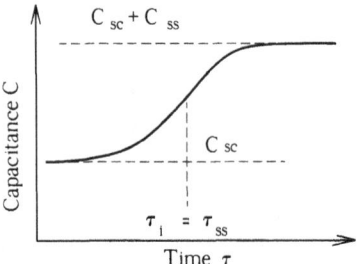

account. This interaction can well distort the results of measurements and may become a source of misunderstanding in their interpretation. This is exemplified in [214, 215], where the authors, in order to interpret their extraordinary experimental findings, have introduced so-called negative surface states. Below we discuss this question in more detail.

A simplified scheme for measuring the impedance at the SE boundary is shown in Figure 8.4. The scheme shows that there are three systems: the system providing the electrode potential φ, the system for measurement of the longitudinal conductance, and the system for measurement of the differential capacitance $C = dQ/d\varphi$. One part of the electric circuit, the SE boundary, is common to them all. The electric perturbation brought about by measurements of the differential capacitance results in an additional potential drop at the SE boundary, $d\varphi$. A value of the potential equal to $\varphi + d\varphi$ will be shown on the potentiostat. Upon comparison of the value of the given potential φ and the measured one, the potentiostat yields the value of $d\varphi$ which is a measure of the incorrectness in the potential φ. This error derived from the potentiostat will be reduced by providing an additional current in the polarization circuit. A decrease of dV

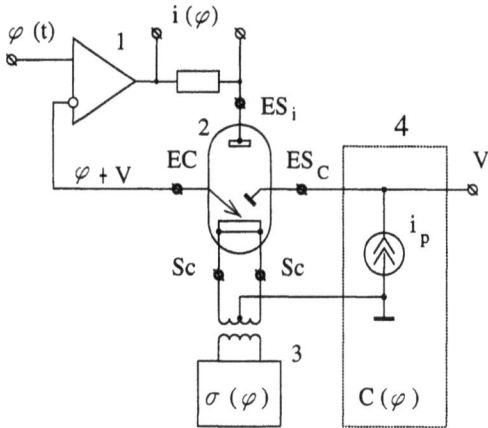

Figure 8.4 Simplified scheme for measuring impedance at the SE interface: (1) potentiostat; (2) electrochemical cell with semiconductor electrode (Sc); (3) scheme of surface conductance measurement; (4) scheme of capacitance measurement (i_p, probe current pulse generator); ES$_i$, subsidiary electrode for polarization; ES$_c$, subsidiary electrode for capacitance measuring; (EC) comparison electrode; $\Delta\varphi$, response of the semiconductor-electrolyte interface; $\varphi(t)$, polarizing voltage; $i(\varphi)$, polarizing current; $(\varphi+\Delta\varphi)$, electrode potential measured in experiment.

at the fixed dQ will, correspondingly, lead to an apparent increase in the measured value of the differential capacitance $C = dQ/d\varphi$. This means that the duration of the test impulse must be much less than the characteristic time of the potentiostat.

One should keep in mind that, in measurements of a semi-conductor surface conductance using FESE, the longitudinal conductance represents a sum of several components

$$\sigma = \sigma_b + \sigma_s + \sigma_{el},$$

where σ_s is the surface conductance and σ_b is the bulk conductance of the semiconductor; σ_{el} is the electrolyte near-electrode layer conductance. The relevant equivalent scheme is shown in Figure 8.5. The part of the current passing through the large Helmholtz layer capacitance increases with increasing frequency. The relevant frequency conductance dispersion is shown in Figure 8.6. The optimal frequency f_h for measurements of surface conductance is below the region corresponding to strong conductance dependence of the frequency (see Fig. 8.6).

139

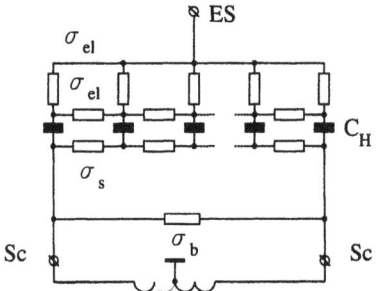

Figure 8.5 Equivalent electric scheme of semiconductor electrode used in surface conductance measurements (see Fig. 8.4).

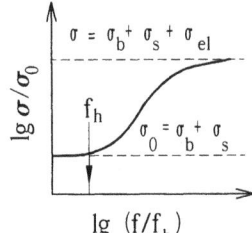

Figure 8.6 Frequency dependence of the surface conductivity, which is the consequency of shunting of semiconductor electrode conductance by the near-surface electrolyte layer.

It is assumed [216] that the frequency range suitable for conductivity measurements is limited from above by the value 40–60 Hz. For the ideally polarizable electrode, the band edges are pinned with respect to the electrolyte with the help of the redox system for arbitrary band bending at the semiconductor surface [17]. In this case there is no potential drop along the electrode surface; hence the shunt current is equal to zero within a wide frequency range. This accounts for the absence of any appreciable frequency dispersion of the longitudinal conductance, observed in a number of experiments [3].

In conclusion, we note that although the FESE technique may be employed in principle in a broad range of semiconductors from wide-gap semiconductors to semimetals, yet, due to the aforementioned shortcomings, its practical realization needs an individual approach in each particular case.

8.2 APPLICATION OF FESE FOR CONTROL OF THE IMPURITY DISTRIBUTION IN SEMICONDUCTORS

Finding the impurity distribution profile is based on measurements of the differential capacitance of the SE interface and/or the surface conductance of a semiconductor in conjunction with a controlled layer-by-layer chemical (electrochemical) etching of the semiconductor. The doping impurity concentration is usually defined using the tangent of the declination angle of the capacitance-voltage characteristic, plotted in Schottky-Mott coordinates [87, 101, 102, 106, 111, 112, 217–222]. The actual distance from the surface (profile) is obtained from the etching rate (during the chemical etching) or from the value of the etching current density (during the electrochemical dissolution of the semiconductor matrix). The details of the etching technique and surface preparation can be found elsewhere (see, for example, [223]).

A systematic study of the possibilities of the FESE for studies of the profile of electrically active impurities in the near-surface region of a semiconductor has been performed in a series of works by Romanov and co-workers [129], Blood [219] and Stevenson and Rodstall [222]. At present, this method and its modifications are widely used in the technology of materials science and microelectronics. It is noteworthy that the application of FESE for control of the doping impurity content and profile is of universal character, that is, it is applicable to a wide range of semiconductors. A disadvantage of this technique is its destructive character.

Some results illustrating the doping impurity profiles for a number of semiconductor materials are presented in Figure 8.7.

8.3   METHOD FOR DETERMINING THE STOICHIOMETRIC COMPOSITION OF SOLID SOLUTIONS

Electrical methods for investigating semiconductors by measuring capacitance-voltage $[C(\varphi)]$ characteristics are widely used to determine the type and concentration of dopants and the spectrum of fast and slow surface states [8, 223]. Additional possibilities in studying semiconductors are opened up when an electrolyte-insulator-semiconductor (EIS) system is used instead of the conventional metal-insulator-semiconductor (MIS) structure. The main advantage of the field-effect technique in the case of the EIS system is the possibility of forming a practically oxide-free surface, with the result that the measured capacitance is in fact the capacitance of the space charge layer of the semiconductor (C_{sc}). Use of specially selected electrolytes and etching regimes yields a surface with a

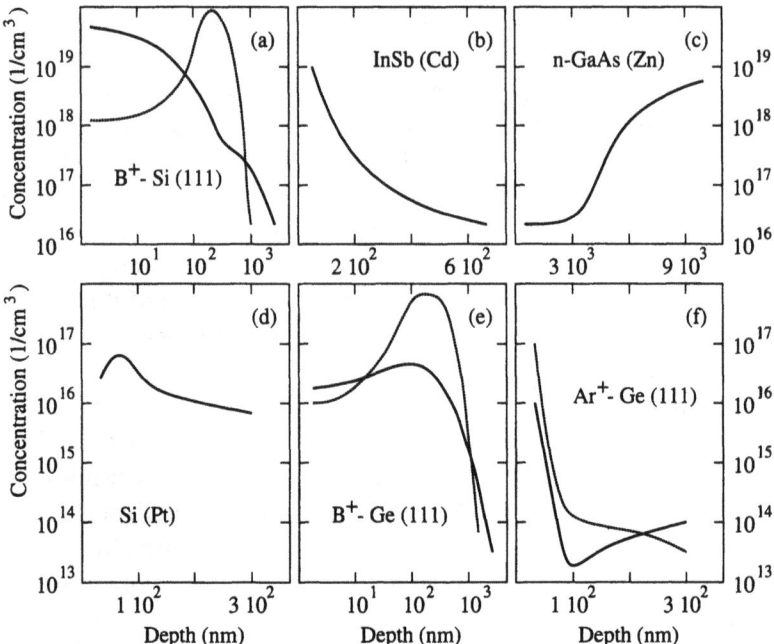

Figure 8.7 Profiles of the doping impurity distribution in the near-surface volume of different semiconductors, obtained from FESE measurement by layer etching: (a), (e), (f) implantation; (b) diffusion; (c), (d) epitaxy; solid line, use of surface conductivity measurement, dashed line, use of differential capacitance measurements.

low density of surface states for various compositions of a mercury cadmium telluride (CMT) solid solution in a wide range of surface potentials.

With the field effect used in electrolytes, a direct comparison of theoretical and experimental $C(\varphi)$ characteristics of CMT is hindered by the unavoidable error in determining the area and roughness factor of the surface of a sample subjected to chemical and electrochemical etching and by the additional capacitance associated with surface states (C_{sc}).

In this Section, a method is proposed for determining the stoichiometric composition of the intrinsic CMT, free of the above experimental errors [204]. The procedure consists in successively taking the logarithm of and differentiating the experimental $C(\varphi)$ characteristics in a wide range of surface potentials, which eliminates the error in determining the area and roughness factor of the surface and allows a more precise comparison with theoretical calculations.

In terms of the one-particle Hartree approximation, the distributions of electron $\rho_e(z)$ and heavy-hole densities $\rho_{hh}(z)$ can be found from a self-consistent solution of the Poisson and Schrödinger equations for electrons and holes[1] (see Section 7.1):

$$\frac{d^2 V(z)}{dz^2} = q \cdot \frac{\rho_e(z) - \rho_{hh}(z) + N_a - N_d}{\varepsilon_0 \varepsilon_{sc}}, \tag{8.1}$$

$$-\frac{d^2 \psi_i(z, k_{\parallel})}{dz^2} = \left\{ k^2[E_i(k_{\parallel}), V(z)] - k_{\parallel}^2 \right\} \psi_i(z, k_{\parallel}), \tag{8.2}$$

$$\left(-\frac{\hbar^2}{2m_{hh}} \frac{d^2}{dz^2} - qV(z) \right) \psi_j(z) = (-E_j - E_g)\psi_j(z). \tag{8.3}$$

We assume that the dispersion law of heavy holes is parabolic and that of electrons is described by the formula (6.1) (see Section 6.1)

$$k^2[E, V(z)] = \frac{1}{P^2} \frac{[E - qV(z)][E - qV(z) + E_g][E - qV(z) + E_g + \Delta]}{[E - qV(z) + E_g + \frac{2}{3}\Delta]}. \tag{8.4}$$

The concentrations of electrons and holes can be found from

$$\rho_e(z) = \frac{1}{\pi} \int_0^{\infty} dk_{\parallel} \sum_i \frac{k_{\parallel} \psi_i(z, k_{\parallel})|^2}{1 + \exp\{[(E_i(k_{\parallel}) - E_F]/(k_0 T)\}}, \tag{8.5}$$

$$\rho_{hh}(z) = \sum_{j=1}^{+\infty} \Gamma_j(E_j)|\psi_j(z)|^2, \tag{8.6}$$

where

$$\Gamma_j(E_j) = \frac{m_{hh} k_0 T}{\pi \hbar^2} \ln\{1 + \exp[(E_j - E_F)/(k_0 T)]\}.$$

The differential capacitance $C_{sc}(V_s)$ is

$$C_{sc} = \frac{dQ_{sc}}{dV_s}, \tag{8.7}$$

---

[1] The light hole density can be neglected to a first approximation.

where $Q_{sc}(V_s) = q \int_0^\infty [\rho_e(z) - \rho_{hh}(z) + N_a - N_d] \, dz$ (in the case in question $N_d = N_a = 0$). By way of example, Figure 8.8 shows a theoretical $C(\varphi)$ characteristic for $Hg_{1-x}Cd_xTe$ with $x = 0.245$. The calculation was done with the following CMT parameters [137, 180] :

$$\varepsilon_{sc} = 20.5 - 15.5x + 5.72x^2, \qquad m_{hh}/m_0 = 0.5, \qquad \Delta = 0.96 \text{ eV},$$

$$P = [(18 - 3x)\hbar^2/(2m_0)]^{1/2},$$

$$E_g = -0.302 + 1.93x + 5.35 \cdot 10^{-4}(1 - 2x)T - 0.81x^2 + 0.832x^3.$$

The employed experimental procedure was based on measuring the impedance of a semiconductor-electrolyte interface probed with a 1-μs-long voltage pulse, with current-voltage characteristics taken simultaneously. The semiconductor electrode was polarized in the potentiostatic regime with a continuous cyclic variation of the electrode potential at rates of $d\varphi/dt = 10$–100 mV/s at $T = 295$ K. The electrode potential (φ) was measured relative to a standard hydrogen electrode. In the range of electrode potentials chosen for polarizing the semiconductor electrode, there were practically no currents across the interface associated with electrochemical reactions, and the field effect had an equilibrium nature.

Prior to measurements, the surface of CMT single crystals was subjected to chemical-dynamical polishing in a bromine-methanol solution. Immediately before an experiment, the sample surface was additionally etched electrochemically. Such a sample treatment removes oxides from the surface and ensures an extremely low density of surface states. As a result, the measured capacitance of the CMT-electrolyte interface is actually the capacitance of the SCL in the semiconductor, that is, $C = C_{sc}$ with the relations $V_s = -(\varphi - \varphi_{fb})$ and $-\Delta\varphi = \Delta V_s$, where φ_{fb} is the flat-band potential, valid over the entire range of electrode potentials.

In this study, the proposed procedure for determining the CMT composition was tested on four single-crystal samples[2] ($x = 0.205$, $0.245, 0.290$, and 0.330) with random crystallographic orientations. Figure 8.8 shows as an example experimental and theoretical $C(\varphi)$ characteristics for the composition with $x = 0.245$. It can be seen that the experimental curve lies somewhat higher than the theoretical

---

[2] Originally, the sample composition was certified using independent Hall effect and optical measurements.

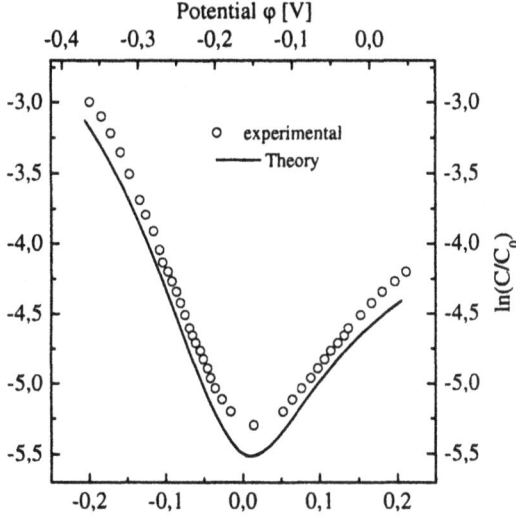

Figure 8.8 Experimental and theoretical $C(\varphi)$ characteristics for a composition with $x = 0.245$.

plot. The best agreement between the theoretical and experimental curves is achieved on making a correction to the sample area by 10%. In this case, the surface state density in the range $-0.1 < V_s < 0.1\,\text{V}$ is estimated to be no more than $3 \cdot 10^{11}$ cm^{-2}. Nevertheless, despite the error in determining the surface area, the proposed procedure allows the stoichiometric composition of the crystal to be determined with sufficient accuracy.

Figure 8.9 presents curves obtained by taking the logarithm and subsequent differentiation of the theoretical and experimental $C(\varphi)$ characteristics ($x = 0.245$). It can be seen that the theoretical (as well as the experimental) dependence of $d\,[\ln(C/C_{sc})]/dV_s$ on V_s shows a minimum in the hole portion ($V_s < 0$) and a maximum in the electron portion ($V_s > 0$). Plotting the derivative values at these extrema as a function of composition (x) yields the curves presented in Figure 8.10. A good agreement between the theory and experiment is seen in the electron portion; a somewhat worse agreement is seen in the hole portion, which is presumably due to uncontrollable oxide growth in anodic polarization.

Thus, the theoretical curves in Figure 8.8 can be used as graphical charts for determining the composition of intrinsic CMT samples. This can be done by taking the logarithm of the experimental $C(\varphi)$

145

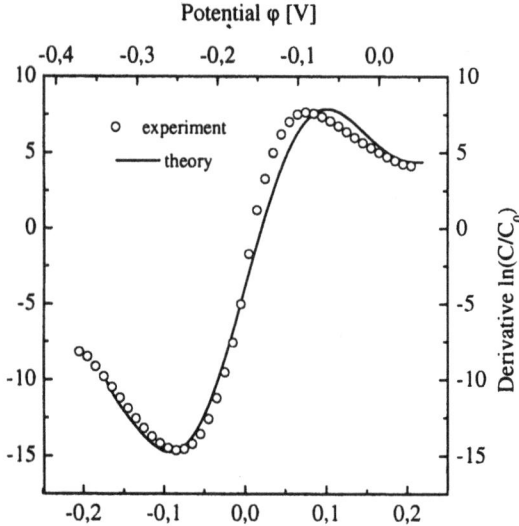

Figure 8.9 Differentiated theoretical and experimental $C(\varphi)$ characteristics $(x = 0.245)$.

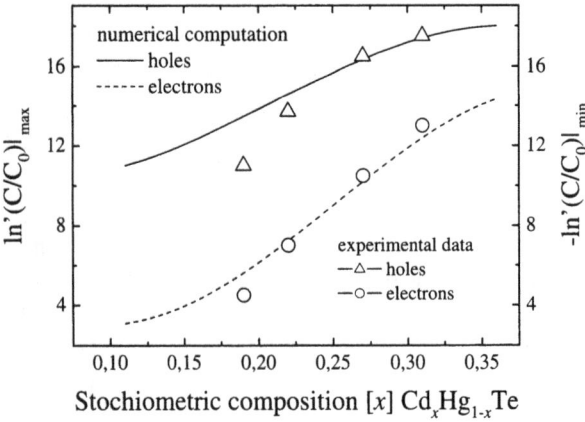

Figure 8.10 Extreme value distribution as a function of composition (x).

characteristic (see Fig. 8.8), differentiating it (see Fig. 8.9), and plotting the obtained minimum and maximum values along the ordinate axis (see Fig. 8.10). The intersection of horizontal lines drawn through these points with the corresponding calculated curves will give the sought-for composition value x. The compositions of test samples determined by this algorithm are listed in the Table 8.1.

TABLE 8.1
Determined Composition (x) of $Hg_{1-x}Cd_xTe$ Solid Solutions

Method of determination				
Nominal value of x	0.205	0.245	0.290	0.330
Found from electron portion x	0.209	0.245	0.295	0.330
Found from hole portion x	0.150	0.255	0.305	0.320

8.4 FESE Technique for Investigation and Control of Biological Systems

Using of the FESE technique for investigation of the nature of biological processes is illustrated in [224], where an electrophysical interpretation of the principles of formation of properties of cell culture samples is presented. A semiconductor electrode with passivated surface is prepared directly in the system under research and is used as a sensor of processes inside the samples. The object of the research is the neoplastic lymphocyte cultures K-562 and NS-0 in appropriate nutrient media.

The kinetic characteristics of differential capacity [$\Delta C(t)$] and electrode potential [$\Delta \varphi(t)$] for semiconductor-bioassay systems are presented in Figure 8.11. Changes in the equilibrium values of appropriate parameters in a stationary condition of the system, which characterize the electrical charge distribution inside the multilayer interface. The nutrient medium plays the role of the electrolyte in the system.

The following facts attract special attention. First, in the absence of live cells (see curves 1 and 2 in Fig. 8.11) the system parameter kinetics have a relaxation character, as opposed to the kinetics of differential capacitance when the bioassay is added [see curve 3 in the $C(t)$ characteristics in Fig. 8.11], which is not monotonic. Second, each system in the absence of live cells has a steady state, shown in the $C(\varphi)$ characteristic and, consequently, the change of electrical charge, taken as a whole, is located on the semiconductor-electrolyte interface. In the presence of a biological culture [see curve 3 in the $C(\varphi)$ characteristics in Fig. 8.11] the steady state of the system is determined not only by the double-layer properties of the electrode interface, but also by an additional capacitance, which is placed beyond the interface and

147

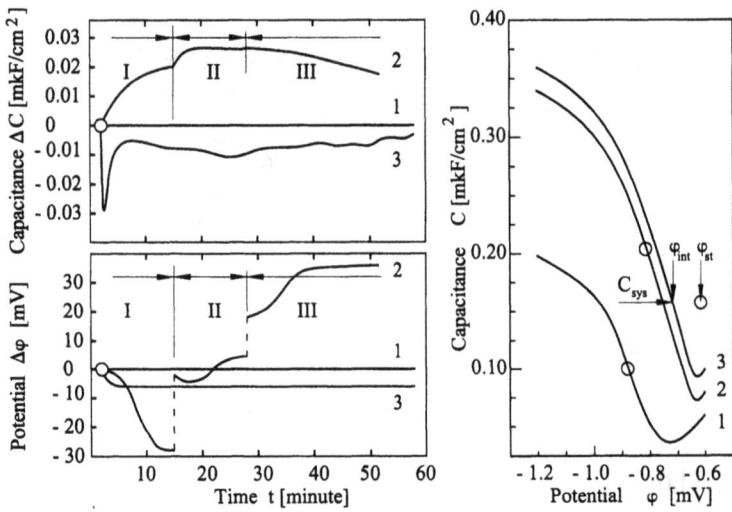

Figure 8.11 Kinetic characteristics of changes of differential capacity $[\Delta C(t) = C(t) - C_{st}]$, electrode potential $[\Delta\varphi(t) = \varphi(t) - \varphi_{st}]$, and $C(\varphi)$ characteristics for several systems, which are composed of a semiconductor electrode and biological samples. Circles designate steady state for each of the systems. Curves correspond to systems with (1), mineral components of the electrolyte; (2) (segment I), sterile nutrient medium; (2) (segment II), addition of a pasteurized portion of K-562 culture to the nutrient medium; (2) (segment III), repeated addition of pasteurized K-562; (3) addition of a live portion of the culture K-562 with volumetric relation of 1:2 to the nutrient medium. Initial concentration of cells in the culture $n \simeq 10^4$ cm^{-3}.

is capable of accumulating an electrical charge. The value of the additional charge (ΔQ) in the system may be estimated as $\Delta Q = (\varphi_{st} - \varphi_{int}) \cdot C_{sys} \approx 16 \cdot 10^{-9}$ C. Considering the average value of the charge located on each live cell as equal to $\delta Q = 3 \cdot 10^{-10}$ C [225, 226] and taking into account the active area of the sensor used ($S = 0.25$ cm^2), one is able to estimate the cell concentration by using the formulas

$$\Delta Q = \delta Q \cdot S \cdot n_\square, \qquad n = \left(\frac{\Delta Q}{\delta Q \cdot S}\right)^{3/2}.$$

This gives $n = n_\square^{3/2} \approx 2.8 \cdot 10^3$ cm^{-3} in the volume of the sample and corresponds to the amount of cells entering the sample during measurement.

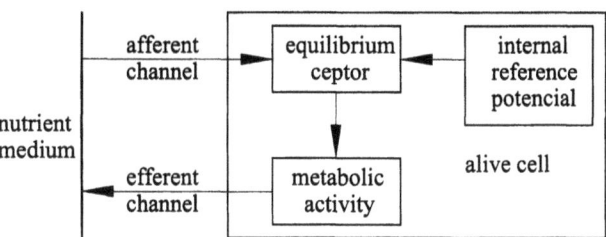

Figure 8.12 Sketch of one cell–nutrient medium system from the point of view of the regulatory model of metabolism.

The observed kinetics of the electrophysical parameters of the system conform well to the basic regulation model for biological systems suggested in [227]. Within the frame of the given model (see Fig. 8.12) the system, by means of the afferent channel, receives data on the electrical potential of the environment, compares them with the appointed value of the internal reference potential, and, depending on the result of the comparison, by means of the efferent channel, exchanges the electrical charge with the external subsystem, which is the environment for each cell. By means of this mechanism each separate cell organizes a uniform subsystem which, in its turn, merges with the external subsystem of the nutrient medium. As a result the biological system of the cell culture is assembled. At the same time the semiconductor sensor constitutes an essential part of the external subsystem giving a new opportunity for control of the biological system as a whole.

Thus, the potentials of both the nutrient medium of the cell culture, which forms an external subsystem for each cell, and the cell culture as a whole is practically constant [see curve 3 in the $\varphi(t)$ characteristic in Fig. 8.11], while the regulatory signal presented as a feedback loop of the system has the maximal amplitude [see curve 3 in the $C(t)$ characteristic in Fig. 8.11], and can be easily registered. Moreover, inasmuch as the semiconductor electrode, used for measurements, is a component of the efferent channel, one can use it for control of the charge exchange process with the external subsystem, governing charge injection, because the peculiarities of the interface allow the injected charge to be changed in both value and sign.

Figure 8.13 depicts the result of control of a system that includes a semiconductor electrode and a biological sample of the cell culture

Figure 8.13 Synchronous kinetics of differential capacitance [$C(t)$] and electrode potential [$\varphi(t)$] of the system after insertion of cell culture NS-0 in nutrient medium at a volumetric ratio 1:2. The cell concentration in the bioassay is $n_0 \approx 10^3$ cm^{-3}. The circles denote steady states of the system in sterile nutrient medium. I and III, inert semiconductor electrode; II, injection semiconductor electrode.

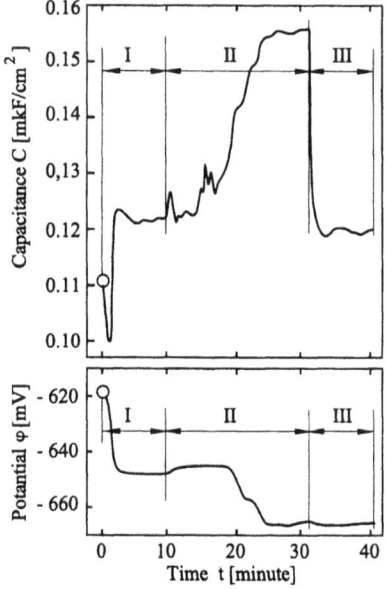

NS-0. The control is carried out by means of a sensor in the external subsystem, which detects the charge properties. The kinetics of differential capacitance [$C(t)$] and electrode potential [$\varphi(t)$] of the system illustrate a transition between two different steady states of the system.

Initially the kinetics of both $C(t)$ and $\varphi(t)$ for the system bioassay NS-0 – inactive electrode (see segment I in Fig. 8.13), qualitatively repeat the peculiarities noted above for the bioassay K-562 [see curves 3 for $\Delta C(t)$ and $\Delta \varphi(t)$ in Fig. 8.11]. During the first 8–10 min the potential of the system and the amplitude of the regulatory efferent signal become adjusted and on condition that external influences are excluded in future they stay constant for an arbitrarily long period.

Segment II in Figure 8.13 corresponds to the system with additional injection of electrical charge from the electrode. This is caused by photogeneration in the range of fundamental absorption of the electrode and results in its depolarization in the range $\Delta \varphi_+ \approx +4$–6 mV. Note in particular the overshoot processes observable in the $C(t)$ characteristic in Figure 8.13 which confirm the feedback charge mechanism by means of the efferent channel. The additional charge participating in the regulatory process can be estimated in recalculation on a single cell as

$$\Delta Q/n_\square = \frac{\Delta \varphi_+ \Delta C}{S \cdot n_0^{(2/3)}} = \frac{2.4 \cdot 10^{-11}}{0.25 \cdot (1/3 \cdot 10^3)^{(2/3)}} \approx 2 \cdot 10^{-13}\, C,$$

which agrees well with coulometric data on transmembrane processes, that is, the whole-cell recording method [225]. Through the initial period, namely, 8–10 min after charge injection, the process of spontaneous electric polarization of the system takes place, and stops during the next 8–10 min. The additional polarization of the system may be estimated as $\Delta \varphi_- \approx -18$–$22\,mV$ and at the same time the excess of charge near the sensor surface increases to $2 \cdot 10^{-8}\,C$. This surplus per cell is given by $\delta Q \approx 4 \cdot 10^{-10}\,C$, which can be interpreted either as a two times increase of charge density on each cell, or as doubling of their number near the control sensor.

Segment III in Figure 8.13 corresponds to the behavior of the system when the additional injection is switched off. Figure 8.13 shows that the electrical potential of the system does not change and the charge excess in the system undergoes fast relaxation with a time constant that is three or four times shorter than that of its initial accumulation. Subsequent switchings of the injection mode of an electrode are satisfied by a system with a fast time constant. This gives an opportunity to govern biological systems by means of feedback control using an external semiconductor electrode.

Thus, the study of charge processes has shown that in biological systems like cell cultures the following processes exist. First, stabilization of the electrical potential in the bulk of the culture is provided by charge exchange between each cell of the culture and its environment; second, the role of the environment for charge exchange can be filled by an external semiconductor electrode; third, management of the electronic properties of an electrode surface allows the parameters of feedback in the system to be changed, and, as a consequence, the status of cell cultures as regulatory systems to be influenced.

◇ Conclusion ◇

$\mathbf{I}$T SHOULD be apparent to the reader at this point that the FESE technique provides a lot of useful information about electrophysical characteristics of semiconductor surfaces [228]. This is accounted for by the strong possibilities offered by this technique for controlled changes and in situ measurements of the surface.

The FESE technique is employed for investigation of a wide variety of materials from insulators to metals. It should be noted, however, that the measured characteristics (the capacitance and conductance) are constituents of the total impedance of a complex SE system. Hence, relevant characteristics can be treated in terms of a conventional FE only if certain conditions are satisfied (see Section 8.1). That is why the potentialities of the FESE are intimately allied to the material investigated and the interface character. A close inspection and careful treatment of the results of field-effect measurements should be performed, taking into account the specific features of the material as well as the processes involved, since the FESE is not only an experimental technique but also a process that forms and affects the SE interphase boundary. In this case, the interpretation of the results implies elaboration of specific physical models reflecting the rearrangements occurring at the interface. It is apparent to the reader that, in many respects, the model and interpretation of results depend on the choice of the corresponding equivalent circuit, which comprises both the investigation scheme and the measuring one and quite often is not unambiguous.

In order to deduce profitable information about physical processes occurring at semiconductor surfaces and the SE interfaces, along with the FESE, a variety of other techniques are employed [22, 229, 230]. These techniques may be significant in themselves and at the same time be supplementary to the FESE technique. Of particular interest is the use of those that, while affecting the SE system, yield its parameters and information on the atomic and molecular level about processes occurring in the near atomic environment. Among the techniques that have proved to be useful in electrode surface studies, one should first of all mention optical ones, including electroluminescence, photoluminescence, ellipsometry, electroreflectance, and so on (see [83, 231–241]). Such techniques as multiple internal attenuated total reflection, Fourier transform infrared spectroscopy, and others [233, 234, 242–248] have considerable promise. The practical significance of these techniques is conditioned on the feasibility of in situ measurements directly in the electrochemical cell [249].

Important information about the atomic structure of the electrode surface and its changes under various perturbations, including polarization and adsorption, may be obtained from X-ray and electron diffraction, X-ray photoelectron, Auger-electron, and synchrotron radation spectroscopy evidence (see [238, 250–252]). Efforts have been made to develop a special electrochemical cell where X-ray measurements can be taken in situ [253, 254], which would provide additional valuable information on the discussed issues that would not otherwise be accessible. A lot is expected from photoinjection, photocurrent, and photoemission techniques [255], and the techniques of modulation spectroscopy [256] and inverse photoemission [257, 258]; scanning tunnel microscopy and atomic force microscopy used directly in electrolytes provide strong potentialities for studies of atomic and electronic structures of electrode surfaces (see [259–266]). We think that techniques which make use of longitudinal and transverse ultrasonic waves and also techniques which employ piezoelectric effects for controlled change of the atomic structure and charge state of the interface may be useful for SE boundary investigations [267, 268]. Information on semiconductor surface properties of the utmost importance may be obtained by making use of an electrolyte with a low freezing temperature and the technique with frozen electrolyte [200, 201, 231, 269–271]; in particular, for investigations of kinetic effects in the conditions of electron gas dimerization near a semiconductor surface, where galvanomagnetic effects studies at low temperatures are needed [200, 201]. Also it may be useful for investigations of electrodes featuring high-temperature superconductivity.

[1]   D. R. Turner. In *The Electrochemistry of Semiconductors*, ed. P. J. Homs. Academ. Press, London, 1962.

[2]   P. J. Boddy. "The Structure of the Semiconductor-Electrolyte Interface (Review)." *J. Electroanal. Chem.* 10:199–244, 1965.

[3]   V. A. Myamlin and Yu. V. Pleskov. *Electrochemistry of Semiconductors.* Plenum Press, New York, 1967.

[4]   H. Gerischer. "Principles of Electrochemistry." In *The CRC handbook of Solid State Electrochemistry*, eds. P. J. Gellings and H. J. M. Bouwmeester. CRC, Boca Raton, FL., 1997.

[5]   H. Gerischer, "Semiconductor Electrochemistry." Chap. 5 in *Physical Chemistry. An Advanced Treatise*, vol. A9 eds. H. Eyring, P. Henderson, and W. Jost. Academic Press. New York, 1970.

[6]   S. Trasatti. "Surface Science and Electrochemistry Concepts and Problems." *Surf. Sci.* 335:1–9, 1995.

[7]   E. M. Stuve, A. Krasnopoler, and D. E. Sauer. "Relating the In-Situ, Ex-Situ and Non-Situ Environments in Surface Electrochemistry." *Surf. Sci.* 335:177–185, 1995.

[8]   A. Many, Y. Goldstein, and N. Grover. *Semiconductor Surfaces.* North-Holland, Amsterdam, 1965.

[9]   J. Dewald. "Semiconductor Electrodes." In *Semiconductors*, ed. N. B. Hannay, Plenum Press, New York, 1959.

[10] S. R. Morrison. *The Chemical Physics of Surfaces.* Plenum Press, New York, 1977.

[11] S. R. Morrison. *Electrochemistry of Semiconductor and Oxidized Metal Electrodes*, Plenum Press, New York, 1980.

[12] R. Memming. "Processes at Semiconductor Electrodes. Chap. 7 in *Comprehensive Treatise on Electrochemistry*, eds. J. O'M. Bockris, Yu. A. Chizmadzhev, B. E. Conway, S. U. M. Khan, S. Sarangapani, S. Stinivasan, R. E. White, and E. Yeager, 529–592. Plenum Press, New York, 1983.

[13] L. M. Peter. "Semiconductor Electrochemistry." Chap. 9 in *Electrochemistry: A Specialist Periodical Report*, eds. G. J. Hills, D. Pletcher, and H. R. Thirsk, 66–100. Burlington House, London, 1984.

[14] P. N. Ross and F. T. Wagner. "The Application of Surface Physics Techniques to the Study of Electrochemical Systems." *Adv. Electrochem. Electrochem. Eng.* 13:69–112, 1984.

[15] J. O'M. Bockris and S. U. M. Khan. *Surface Electrochemistry.* Plenum Press, New York, 1993.

155

[16] Yu. V. Pleskov and Yu. Ya. Gurevich. "Electrochemistry of Semi-conductors: New Problems and Prospects." *Mod. Aspects Electrochem.* 16: 189–251, 1985.

[17] Yu. V. Pleskov and Yu. Ya. Gurevich. *Semiconductor Photoelectrochemistry.* Consultants Bureau, New York, 1986.

[18] K. Uosaki and H. Kita. "Theoretical Aspects of Semiconductor Electrochemistry." *Mod. aspects electrochem.* 18:1–60, 1986.

[19] A. K. Vijh. "Perspectives in Electrochemical Physics." *Mod. Aspects Electrochem.* 17:1–30, 1986.

[20] A. Hamnett. "Semiconductor Electrochemistry." *Compr. Chem. Kinet.* 27:61–246, 1987.

[21] H. Finklea, ed. *Semiconductor Electrodes.* Elsevier, Amsterdam, 1988.

[22] *Modern Aspects of Electrochemistry*, ed. J. O'M. Bockris, B. E. Conway, and R. E. White. Plenum Press, New York, 1989–1992.

[23] R. D. Rauch. "Photoelectrochemical Processing of Semiconductors and Electronics." In *Electrochemistry of Semiconductors and Electronics*, eds. J. McHardy and F. Ludwig, vol. 4, 177–216. Noyes Publications, Park Ridge, N.Y., 1992.

[24] T. Furtak. "Electrochemical Surface Science." *Surf. Sci.* 293:945–955, 1994.

[25] Yu. V. Pleskov. "Semiconductor Photoelectrochemistry for Cleaner Environment: Utilisation of Solar Energy." In *Environmental Oriented Electrochemistry*, ed. C. A. C. Sequeira, 417–443, Elsevier, New York, 1994.

[26] M. X. Tan, P. E. Laibinis, S. T. Nguyen, J. M. Kesselman, C. E. Stanton, and N. S. Levis. "Principles and Applications of Semiconductor Photoelectrochemistry." In *Principles and Applications of Semiconductor Photoelectrochemistry, Progress in Inorganic Chemistry.* ed. K. D. Karlin, vol. 41, 21–144. Wiley, New York, 1994.

[27] W. Jaegermann. "The Semiconductor/Electrolyte Interface: A Surface Science Approach." *Mod. Aspects Electrochem.* 30:1–185, 1996.

[28] J. McHardy and F. Ludwig, eds. *Electrochemistry of Semiconductors and Electronics.* Noyes Publications, Park Ride, N.Y., 1992.

[29] W. P. Gomes and H. H. Goossens. "Electrochemistry of III-V Compound Semiconductors: Dissolution Kinetics and Etching." *Adv. Electrochem. Eng.* 3:1–54, 1994.

[30] S. Licht, O. Khaselev, T. Soga, and M. Umeno. "Multiple Bandgap Photoelectrochemistry: Energetic Configurations for Solar Energy Conversion." *Electrochem. Solid State Lett.* 1:20–23, 1998.

[31] H. Gerischer. "Principles of Electrochemistry (Solid State)." In *The CRC Handbook of Solid State Electrochemistry*, ed. P. J. Gellings and H. J. M. Bouwmeester, 9–73. CRC, Boca Raton FL, 1997.

[32] I. Rubinstein. "Fundamentals of physical electrochemistry." In *Physical Electrochemistry: Principles, Methods and Applications*, ed. I. Rubinstein, 1–25. Marcel Dekker, New York, 1995.

[33] Yu. V. Pleskov. "Solar Energy Conversion in Photoelectrochemical Cells with Semiconductor Electrodes." *Prog. Surf. Sci.* 15:401–456, 1984.

[34] W. J. Albery and A. W. Foulds. "Photogalvanic Cells." *J. Photochem.* 10:41–57, 1979.

[35] H. Gerischer. "Photodecomposition of Semiconductors: Thermodynamic, Kinetics and Application to Solar Cells." *Faraday Discuss. Chem. Soc.* 70:137–154, 1980.

[36] P. Salvador, N. Aloso-Vante, and H. Tributsch. "Photoelectrocatalytic Study of Water Oxidation at n-RuS_2 Electrodes." *J. Electrochem. Soc.* 145:216–241, 1998.

[37] R. Memming. "Photoelectrochemical Solar Energy Conversion Electrochemistry." In *Electrochemistry*, ed. E. Steckhan, vol. 2, 79–112. Springer, New York, 1988.

[38] W. Brattain, and C. Garrett. "The Investigation of the Germanium-Electrolyte Interfaces." *Bell Syst. Tech. J.* 34:129–134, 1955.

[39] P. P. Konorov and M. N. Kolbin. "Research of Change of the Diffusion Length of Displacement of Carriers of a Current and Electrode Potential of germanium from Electrolyting to Processing." *Fiz. Tverd. Tela* 3:1553–1556, 1961 (in Russian).

[40] P. P. Konorov and O. V. Pomanov. "Current-Voltage Characteristics on Germanium-Electrolyte Interface." *Vestn. Leningr. Univ.* 22:65–70, 1965 (in Russian).

[41] P. P. Konorov. About Equilibrium and Nonequilibrium Field-Effect in Semiconductor on Border with Electrolyte." *Vestn. Leningr. Univ.* 10:50–52, 1972 (in Russian).

[42] O. V. Romanov and P. P. Konorov. "The Field Effect and Surface States on the Germanium-Electrolyte Interface." *Fiz. Tverd. Tela* 8:13–20, 1966 (in Russian).

[43] O. V. Romanov. "Nature Cristallo-Physico-Chimique de l'Interface entre un Semiconducteur et Son Oxyde Propre. I. Etudes Experimentales des Semiconducteurs des Groupes A^4 et A^3B^5." *Rev. Phys. Appl.* 19:379–388, 1984.

[44] O. V. Romanov. "Nature Cristallo-Physico-Chimique de l'Interface entre un Semiconducteur et Son Oxyde Propre. II. Une Conception Générale de l'Interface Semiconducteur-Oxyde Propre." *Rev. Phys. Appl.* 19:389–394, 1984.

[45] G. Nicolis and I. Prigogine. *Exploring Complexity. An Introduction.* W. H. Freeman and Company, New York, 1990.

[46] R. Memming. "Photoelectrochemical Solar Energy Conversion." *Electroanal. Chem.* 11:79–112, 1979.

[47] R. Wilson. "Electron Transfer Processes at the Semiconductor-Electrolyte Interface." *CRC Crit. Rev. Solid State Mater. Sci.* 10:1–117, 1980.

[48] R. Williams. "Electrochemical Reactions of Semiconductors." *J. Vac. Sci. Technol.* 13:12–24, 1976.

[49] K. Sangval. *Etching of Crystals: Theory, Experiment and Application.* North-Holland, Amsterdam, 1987.

[50] L. Blum. "Structure of the Electrical Double Layer." *Adv. Chem. Phys.* 78:171–222, 1990.

[51] P. Attard. "Electrolytes and the Electric Double Layer." *Adv. Chem. Phys.* 92:1–159, 1996.

[52] P. Delahay. *New Instrumental Methods in Electrochemistry.* Interscience, New York, 1958.

[53] R. Marcus. "Chemical and Electrochemical Electron-Transfer Theory." *Annu. Rev. Phys. Chem.* 15:174–180, 1964.

[54] K. Seeger. *Semiconductor Physics.* Springer-Verlag, Vienna, 1973.

[55] S. M. Sze. *"Physics of Semiconductor Devices.* 2nd ed. Wiley-Interscience, New York, 1981.

[56] G. G. Kareva and P. P. Konorov. "About the Polarizing Characteristics of *n*-Si in Alkali Solution." *Elektrokimiya*, 6:121–122, 1970 (in Russian).

[57] P. P. Dogonadze and A. M. Kuznetzov. "The Kinetic of the Heterogenic Chemical Reactions in Solutions." *Kinet. katal. Itogi Nauk. Techn.* 5:223–300, 1978 (in Russian).

[58] A. Modinos. *Field, Thermoionic and Secondary Electron Emission Spectroscopy.* Plenum Press, New York, 1984.

[59] G. G. Kareva. "The Electrochemical Properties of the Si at Cathodic Polarization." *Elektrokimiya*, 19:1671–1674, 1983 (in Russian).

[60] P. P. Konorov, V. N. Shvetsov, and S. V. Schegolihina. "Effects of the Nonequilibrium Depletion in Germanium at the Electrolyte Interface." *Phys. Status Solidi*, 30:845–850, 1968.

[61] A. M. Yafyasov, V. V. Monachov, and O. V. Romanov. "The Spectroscopy of the Surface States by the Method of Field Effect in Electrolytes." *Vestn. Leningr. Univ.* 1:104–107, 1986 (in Russian).

[62] W. Mönch. *Semiconductor Surfaces and Interfaces*, ed. G. Ertl, R. Goneer, and D. L. Mills. Springer-Verlag, Berlin, 1993.

[63] T. Sugano, K. Hoh, and H. Sakaki. "Quantum State and Electron Transport at Silicon–Silicon Dioxide Interface. *J. Fac. Eng. Univ. Tokyo (B)* 32:178–246, 1973.

[64] B. Gergel and R. Suris. "The Theory of the Surface States and Conductivity in MIS-Structures." *Sov. Phys. JETP* 84:719–730, 1983.

[65] J. Siugh and A. Madhuner. "Origin of U-Shaped Background Density of Interface States at Nonlattice Matched Semiconductor Interface." *J. Vac. Sci. Technol.* 19:437–442, 1981.

[66] V. N. Ovsiuk. *Electronic Properties of Semiconductors with Space Charge Region.* Nauka, Novosibirsk, 1984 (in Russian).

[67] H. Gatos. In *Surface Chemistry of Metals and Semiconductor,* ed. H. C. Gatos. Wiley, New York, 1960.

[68] M. Cretella and H. Gatos. "The Reaction of Germanium with Nitric Acid Solutions." *J. Electrochem. Soc.* 105:487–495, 1958.

[69] P. P. Konorov, O. V. Romanov, and G. G. Kareva. "Research on the Surface States Arising during Oxidation of Germanium. *Fiz. Tverd. Tela* 8:2517–2519, 1966 (in Russian).

[70] P. Boddy and W. Brattain. "Effect of Cupric Ion on the Electrical Properties on the Germanium–Aqueous Electrolyte Interface. *J. Electrochem. Soc.* 109:812–818, 1962.

[71] O. V. Romanov, Yu. V. Demashev, and A. D. Andreev. "Influence of Adsorption of Fe, Co and Ni Ions on Germanium Surface Properties." *Fiz. Tekh. Poluprovodn.* 4:1335–1337, 1970 (in Russian).

[72] O. V. Romanov and A. D. Andreev. "The Surface States Arising at Rh Ion Adsorption on Germanium Surfaces." *Fiz. Tekh. Poluprovodn.* 11:2116–2125, 1973 (in Russian).

[73] O. V. Romanov and M. Ribeiro de Silva. "The Electrical Actions of Au on Germanium Surface." *Mikroelektronika* 9:494–498, 1975 (in Russian).

[74] B. I. Boltaks. *Diffusion in Semiconductors.* Academic Press, New York, 1963.

[75] V. E. Primachenko, O. V. Snitko, and V. V. Milenin. "Non-Equilibrium Field Effect on Si in the Region of High Depletion." *Phys. Status Solidi,* 11:711–718, 1965.

[76] N. Harrick. "Semiconductor Surface Properties Deduced from Free Carrier Absorption and Reflection of Infrared Radiation." *J Phys. Chem. Solids* 14:60–71, 1960.

[77] P. P. Konorov and S. V. Schegolihina. "Modulation of Light Absorption on the Germanium-Electrolyte Interface." *Fiz. Tverd. Tela* 9:2117–2119, 1967 (in Russian).

[78] R. Williams. "High Electric Fields in Cadmium Sulfide. Field Effect Constriction of Current Flow and Dielectric Breakdown." *Phys. Rev.* 123:1646–1651, 1961.

[79] R. Williams. "Dielectric Breakdown in Cadmium Sulfide." *Phys. Rev.* 125:850–855, 1962.

[80] P. P. Konorov, Yu. A. Tarantov, and A. I. Gurjev. "Non-Equilibrium and Surface Barrier Photo-Effect on the GaAs-Electrolyte Interface. *Vestn. Leningr. Univ. Fiz., Khim.* 10:47–52, 1973 (in Russian).

[81] G. G. Kareva and P. P. Konorov. "Photoconductivity of Germanium and Silicon in the High Condition Surface Band Bending." *Fiz. Tekh. Poluprovodn.* 6:271–275, 1972 (in Russian).

[82] H. De-Vore. "Spectral Distribution of Photoconductivity." *Phys. Rev.* 102:86–91, 1956.

[83] K. Shaklee, F. Pollak, and M. Cardona. "Electroreflectance at a Semi-conductor-Electrolyte Interface." *Phys. Rev. Lett.* 15:883–885, 1965.

[84] C. Wilmsen and S. Szpak. "MOS-Processing for A^3B^5 Compounds." *Thin Solid Films* 46:17–45, 1977.

[85] C. Wilmsen. "Oxide Layers on III-V Compound Semiconductors." *Thin Solid Films* 39:105–117, 1976.

[86] C. Wilmsen. "Chemical Composition and Formation of Thermal and Anodic Oxide III-V Compound Semiconductor Interfaces." *J. Vac. Sci. Technol.* 19:279–289, 1981.

[87] K. P. Quinlan, P. W. Yip, A. K. Rai, and T. N. Wittberg. "Oxidation of n-InP and Indium in the Negative Potential Region at pH5." *J. Electrochem. Soc.* 143:524–530, 1966.

[88] I. M. Tsidilkovski. *Electron Spectrum of Gapless Semiconductors.* Springer-Verlag, Berlin, 1997.

[89] T. Sawada and H. Hasegawa. "Interface State Band between GaAs and Its Anodic Native Oxide." *Thin. Solid Films* 56:183–187, 1979.

[90] M. Nishida. "Energy Distribution of Dangling Orbital Surface States on the (110) surface of III-V Compounds." *Phys. Status Solidi B* 90:K39–K43, 1980.

[91] N. L. Dmitruk and V. G. Litovchenko. "Investigation of the Photoactive Surface States in GaAs by IR-Spectroscopy." *Fiz. Tekh. Poluprovodn.* 14:1555–1563, 1980 (in Russian).

[92] C. Hoffman, H. Gerritsen, and A. Nurmikko. "Study of Surface Recombination in GaAs and InP by Picosecond Optical Techniques." *J. Appl. Phys.* 51:1603–1604, 1980.

[93] W. E. Spicer, I. Lindau, P. R. Skeath, and C. Y. Su. "Unified Defect Model and Beyond." *J. Vac. Sci. Technol.* 17:1019–1027, 1980.

[94] Fan Fu-Pen and J. Bard. "Semiconductor Electrodes. 24. Behavior and Photo-Electrochemical Cells Based on p-Type GaAs in Aqueous Solution." *J. Am. Chem. Soc.* 102:3677–3683, 1980.

[95] K. Frese and S. Morrison. "Electrochemical Measurements of Interface States at GaAs/Oxide Interface." *J. Electrochem. Soc.* 126:1235–1241, 1979.

[96] A. Wolkenberg. "Measurements of Physical Properties of GaAs and Si by Electrochemical Methods." *Surf. Sci.* 50:580–590, 1975.

[97] W. Laflere, F. Cardon, and W. Gomes. "On the Differential Capacitance of the n and p-Type Gallium Arsenide Electrode." *Surf. Sci.* 44:541–548, 1974.

[98] P. Janietz, R. Weiche, and J. Westfahl. "On Capacity Measurements and the Energy Distribution of Surface States at the Electrolyte-GaAs Interface." *J. Electroanal. Chem.* 106:23–33, 1980.

[99] T. Ambridge and M. Faktor. "Electrochemical Capacitance Characterization of n-type GaAs." *J. Appl. Electrochem.* 4:135–143, 1974.

[100] E. A. Meulenkamp and A. R. de Wit. "Suppression of Hole Investigation from Ce^{4+} and $Fe(CN)_6^{3-}$ at p-type semiconductors by illumination." *J. Electrochem. Soc.* 141(1):109–115, 1996.

[101] A. M. Gancalves, C. Mathieu, M. Herlem, and A. Etchebery. "Uncommon Behaviour of the p-GaAs Electrode during the Reduction of Oxygen in Liquid Acidic Ammonia." *J. Electroanal. Chem.* 420: 25–29, 1997.

[102] P. M. Vereecken, F. Vanden Kerchove, and W. P. Gomes. "Electrochemical Behaviour of (100) GaAs in Copper-Containing Solution." *Electrochim. Acta* 41:95–107, 1996.

[103] O. V. Romanov. "Potentiodynamic Investigation on Semiconductor Electrodes: Semiconductors A_3B_5. *Elektrokhimiya* 19:153–160, 1983 (in Russian).

[104] O. V. Romanov and V. B. Bogevolnov. "The Field-Effect on GaP_xAs_{1-x} Surfaces." *Fiz. Tekh. Poluprovodn.* 18:1408–1412, 1984 (in Russian).

[105] B. H. Erne, D. Vanmaekelbergh, and J. J. Kelly. "Morphology and Strongly Enhanced Photoresponse of GaP Electrodes Made Porous by Anodic Etching." *J. Electrochem. Soc.* 143:305–314, 1996.

[106] Sh. Koha, M. Peterson, D. Arent, J. Redwing, M. Tischler, and J. Turner. "Electrochemical Investigation of the Gallium Nitride–Aqueous Electrolyte Interface." *J. Electrochem. Soc.* 142:L238–L240, 1995.

[107] C.-J. Hung, L. Halaoui, A. Bard, P. Grudowski, R. Dupuis, J. Molstad, and F. DiSalvo. "Electroluminescence at GaN and $Ga_xIn_{1-x}N$ Electrodes in Aqueous Electrolytes." *Electrochem. Solid State Lett.* 1:142–144, 1998.

[108] A. M. Yafyasov and V. B. Bogevolnov. "Investigation of Semiconductor-Electrolyte Interface in High Electric Fields. *Photoelectronika* 3:111, 1990 (in Russian).

[109] *Display Devices*, ed. J. I. Pankov. Springer-Verlag, Berlin, 1980.

[110] O. V. Romanov, A. V. Popov, and M. A. Sokolov. "Potentiodynamic Investigation on InP-Semiconductor Electrodes." *Elektrokhimiya* 20: 1086–1092, 1984 (in Russian).

[111] I. E. Vermeir and W. P. Gomes. "The Etching of InP by Acidic Iodine Solutions." *J. Electrochem. Soc.* 143:1319–1325, 1996.

[112] J. Schefold. "Impedance and Current-Voltage Behavior of a Semiconductor/Liquid Junction with a Small Energy Barrier." *J. Electrochem. Soc.* 143:1598–1603, 1996.

[113] H. Law. "Anodic Oxidation of InGaAsP." *Appl. Phys. Lett.* 37:68–70, 1980.

[114] V. V. Monachov, A. M. Yafyasov, and O. V. Romanov. "Determination of Density States in InSb and PbS Bands." *Sov. Phys. Semicond.* 20:603–608, 1986.

[115] O. V. Romanov and A. V. Popov. "Potentiodynamic Investigation on InAs-Semiconductor Electrodes." *Elektrokhimiya* 20:921–928, 1984 (in Russian).

[116] I. M. Tsidil'kovskii. *Band Structure of Semiconductors*. Pergamon Press, Oxford, 1982.

[117] H. Meincke, D. Ebling, J. Heinze, M. Tacke, and H. Bottner. "Potentiostatic Oxide Formation on Lead Selenide Single Crystals in Alkaline Solutions." *J. Electrochem. Soc.* 145:2806–2813, 1998.

[118] Yu. L. Michlin and E. B. Tomashevich. "Impedance of PbS Electrode in Water Solution." *Elektrokhimiya* 26:607–613, 1990 (in Russian).

[119] M. G. Danilova, L. L. Sveshnikova, and S. M. Repinsky. "The Comparative Investigation of Electrochemical Properties of Pb, Te and PbTe." *Elektrokhimiya* 26:788–790, 1990 (in Russian).

[120] A. M. Yafyasov. "Electrical Properties of PbTe and (PbSn)Te Surfaces." *Semiconductors* 28(4):371–373, 1994.

[121] M. Cohen and Y. Tsang. "Theory of the Electronic Structure of Some IV-VI Semiconductors." *J. Phys. Chem. Soc.*, suppl. 32:303–317, 1972.

[122] J. Dimmock. "K-p Theory for the Conduction and Valence Bands of $Pb_{1-x}Sn_xTe$ and $Pb_{1-x}Sn_xSe$ Alloys." *J. Phys. Chem. Soc.*, suppl. 32:319–330, 1972.

[123] A. Anagnostopoulos, V. B. Bogevolnov, I. M. Ivankiv, O. Yu. Shevchenko, A. D. Perepelkin, and A. V. Yafyasov. "The Electrophysical Properties on the Surface Layer on the Semiconductor $TlBiSe_2$." *Phys. Status Solidi B* 231(2):451–456, 2002.

[124] O. Yu. Shevchenko, A. M. Yafyasov, V. B. Bogevolnov, I. M. Ivankiv, and A. D. Perepelkin. "Field Effect in System Consisting of Electrolyte and $(TlBiSe_2)_{1-x}$-$(TlBiS_2)_x$ Solid Solution." *Semiconductors* 36(4): 420–423, 2002.

[125] M. Ozer, K. M. Paraskevopoulos, A. N. Anagnostopoulos, S. Kokkou, and e. K. Polychroniadis. "Single Crystal Growth and Characterization of Narrow-Gap $(TlBiSe_2)_{1-x}$-$(TlBiS_2)_x$ Mixed Crystals." *Semicond. Sci. Technol.* 13: 86–90, 1998.

[126] C. l. Mitsas, E. K. Polychroniadis, and D. I. Siapkas. "Characterization of the Interfacial Region of Epitaxial $TlBiSe_2$ Thin Films by Infrared Spectroscopy and Transmission Electron Microscopy." *Thin Solid Films* 353:85–91, 1999.

[127] A. M. Yafyasov, V. B. Bogevolnov, and A. D. Perepelkin. "Electrophysical Properties of Layered Structure on a Basis (CdHg)Te in Semiconductor-Electrolyte Systems." *Sov. Phys. Semicond.* 25:1339–1344, 1991.

[128] A. M. Yafyasov, A. D. Perepelkin, and V. B. Bogevolnov. "Investigation of the Parameters of the Energy Band Structure of Surface Layers of Zero-Gap Semiconductors (CdHg)Te and HgTe by the Field

Effect Method in Electrolytes." *Sov. Phys. Semicond.* 26(4):360–364, 1992.

[129] O. V. Romanov, A. M. Belych, E. L. Krasnov, and V. I. Kalenic. Electronic Properties of Surface, Subsurface Layers and Interfaces for Semiconductor and Semimetal Solid Solutions." *Surf. Sci.* 269:1032–1036, 1992.

[130] C. Gabrielli. Chap. 6 in *Physical Electrochemistry: Principles, Methods and Applications*, ed. I. Rubinstein, 243–292. Marcel Dekker, New York, 1995.

[131] R. K. Willardson and A. C. Beer, eds. *Semiconductors and semimetals: Mercury Cadmium Telluride.* Academic Press, New York, 1981.

[132] D. R. Rhiger and R. E. Kvaas. "Composition of Native Oxides and Etched Surfaces on $Hg_{1-x}Cd_xTe$." *J. Vac. Sci. Technol.* 21:168–171, 1982.

[133] U. Solzbach and H. J. Richter. "Sputter Cleaning and Dry Oxidation of CdTe, HgTe, and $Hg_{0.8}Cd_{0.2}Te$ Surfaces." *Surf. Sci.* 97:191–205, 1980.

[134] S. P. Kowalchik and J. T. Cheung. "Native Oxides on CdHgTe." *J. Vac. Sci. Technol.* 18:944–953, 1981.

[135] A. Lastras-Martinez and U. Lee. "Electrolyte Electroreflectance Study of the Effects of Anodization and of Chemo-Mechanical Polish on $Hg_{1-x}Cd_xTe$." *J. Vac. Sci. Technol.* 21:157–163, 1982.

[136] A. M. Yafyasov, V. G. Savitski, R. N. Kovtun, A. D. Perepelkin, and V. B. Bogevolnov. "The Band Structure Parameters of Surface Layers of Epitaxial Films of Narrow-Gap Solid Solutions." *Semiconductors* 27(5):419–422, 1993.

[137] N. N. Berchenko, V. E. Krevs, and V. G. Sredin. *Semiconductor Solid Solutions and Their Applications, Reference Tables.* Voenizdat, Moscow, 1982.

[138] A. M. Yafyasov, A. D. Perepelkin, Yu. N. Myasoedov, and M. I. Matveev. "Electrophysical Properties of the (MnHg)Te Surfaces." *Fiz. Tekh. Poluprovodn.* 24:875–878, 1990 (in Russian).

[139] B. L. Gelmont, V. I. Ivanov-Omskij, and I. M. Tzidilkovsky. "The Electron Energy Spectrum on Gapless Semiconductors." *Usp. Fiz. Nauk* 120:337–362, 1976 (in Russian).

[140] N. N. Berchenko and M. T. Pashkovski. "Telluride Mercury Semiconductor with Zero Band Gap." *Usp. Fiz. Nauk* 119:223–255, 1976 (in Russian).

[141] O. V. Romanov, V. B. Bogevolnov, and Yu. N. Myasoedov. "The Field-Effect on HgTe and $Cd_{0.2}Hg_{0.8}Te$ Surfaces." *Sov. Phys. Semicond.* 18:661–665, 1987.

[142] A. M. Yafyasov, V. B. Bogevolnov, and A. D. Perepelkin. The Field-Effect on Gapless Semiconductors." *Sov. Phys. Semicond.* 21:697–703, 1987.

[143] A. M. Yafyasov, V. B. Bogevolnov, and A. D. Perepelkin. "Parameters of Band Structure in Surface Layers of Zero-Gap Semiconductors (CdHg)Te." *Phys. Status Solidi B* 183:419–423, 1994.

[144] P. C. Janovitz, N. Orlowski, R. Manzke, and Z. Golacki. "On the Band Structure of HgTe and HgSe: View from Photoemission. *J Alloys Compounds* 328:84–89, 2001.

[145] K.-U. Gawlik, L. Kipp, M. Skibowski, N. Orlowski, and R. Manzke. "HgSe: Metal or Semiconductor?" *Phys. Rev. Lett.* 78:3165–3170, 1997.

[146] M. von Truchsess, A. Pfeuffer-Jeschke, C. R. Becker, G. Landwehr, and E. Batke. "Electronic Band Structure of HgSe from Fourier Transform Spectroscopy. *Phys. Rev. B* 61:1666–1669, 2000.

[147] D. Eich, D. Huåbner, R. Fink, E. Ubach, K. Ortner, C. R. Becker, G. Landwehr, and A. Flesar. "Electronic Structure of HgSe (001) Investigated by Direct and Inverse Photoemission. *Phys. Rev. B* 61:12666–12669, 2000.

[148] I. Stolpe, O. Portugall, N. Puhlmann, H.-U. Mueller, M. von Ortenberg, M. Von Truchsess, C. R. Becker, A. Pfeuffer-Jeschke, and G. Landwehr. "Intra- and Inter-Band Transition in HgSe in Megagauss Fields." *Physica B* 294/295:459–462, 2001.

[149] O. Yu. Shevchenko, V. F. Radantsev, A. V. Yafyasov, V. B. Bogevolnov, I. M. Ivankiv, and F. L. Perepelkin. "Determination of the Matrix Element of the Quasi-Momentum Operator in the Zero-Gap Semiconductor HgSe by the Field-Effect Method in an Electrolyte." *Semiconductors* 36:390–393, 2002.

[150] H. Gerischer. "An Interpretation of the Double Layer Capacity of Graphite Electrodes in Relation to the Density of States at the Fermi Level." *J. Phys. Chem.* 89:4249–4251, 1985.

[151] A. M. Yafyasov, V. B. Bogevolnov, and S. P. Zelenin. "The Differential Capacitance of the Carbon-Electrolyte Interface." *Vestn. Leningr. Univ., Fiz., Khim.* 91–93, 1989 (in Russian).

[152] A. M. Yafyasov, V. B. Bogevolnov, and S. P. Zelenin. "Manifestation of the Space Charge Region Size Quantization on Differential Capacitance and Surface Conductivity of the Graphite Electrodes." *Elektrokhimiya* 25:536–538, 1989 (in Russian).

[153] R. Noziers. "Cyclotron Resonance in Graphite." *Phys. Rev.* 109: 1510–1515, 1958.

[154] R. C. Tatar and S. Rabai. "Electronic Properties of Graphite: A Unified Theoretical Study." *Phys. Rev. B* 25:4126–4141, 1982.

[155] Z. Galus. "Mercury Electrodes. Chap. 14 in *Laboratory Techniques in Electroanalytical Chemistry*, ed. P. T. Kissinger and W. R. Heineman, 469–485. Dekker, New York, 1996.

[156] A. P. Frumkin. *The Potential of Zero Charge.* Nauka, Moscow, 1979 (in Russian).

[157] J. P. Wadisli. "Contribution of the Metal to the Differential Capacitance of the Ideally Polarizable Electrode." *Electrochim. Acta* 124:149–154, 1986.

[158] N. F. Mott. *Metal-Insulator Transitions*. Taylor and Francis, London, 1974.

[159] J. T. McDevitt, D. R. Riley, and S. G. Haupt. "Electrochemistry of High Temperature Superconductors: Challenges and Opportunities." *J. Anal. Chem.* 65:535A–545A, 1993.

[160] O. A. Petri and G. A. Tsirlina. "Electrochemistry of Oxide High-Temperature Superconductors." *Adv. Electrochem. Sci. Electrochem. Eng.* 5:61–123, 1997.

[161] J. T. McDevitt, S. G. Haupt, and C. E. Jones. "Electrochemistry of High-T_c Superconductors." *Electroanal. Chem.* 19:337–486, 1996.

[162] K. Kaneto and K. Yoshino. "Characteristics and Application of Superconducting Ceramics as Substrate in Electrochemical Processes." *Tech Report, Osaka Univ.* 38:97–101, 1988.

[163] H. Bachtler and W. J. Lorenz. "Electrochemical Behaviour of $YBa_2Cu_3O_7$." *J. Electrochem. Soc.* 135:2284–2287, 1988.

[164] H. R. Khan and Ch. J. Raub. "Degradation of Superconductivity in the Superconducting Oxide $YBa_2Cu_3O_{7-x}$ in Aqueous Solutions." *J. Less. Common Met.* 146:L1–L5, 1989.

[165] M. T. San Jose and A. M. Espinosa. "Electrochemical Behaviour of Copper Oxides at a Carbon Paste Electrode: Application to the Study of the Superconductor YBaCuO." *Electrochim. Acta* 36:1209–1218, 1991.

[166] K. Shao, C. H. CaO, H. Zhy, H. Y. Shen, and T. J. Li. "Investigation of Photolithography for Fabricating YBaCuO Superconducting Thin Film Devices with Chemical Etching." *Mod. Phys. Lett.* 1:375–381, 1988.

[167] L. B. Harris and F. K. Nyang. "Stability of Yttrium Based Superconductors in Moist Air." *Solid State Commun.* 67:359–362, 1988.

[168] S. E. Triler, S. D. Atkinson, and P. A. Tuiezer. "Dissolution of $YBa_2Cu_3O_{7-x}$ in Various Solutions." *Am. Ceram. Soc. Bull.* 67:759–762, 1988.

[169] Yu. S. Popov, O. N. Gorshkov, and E. S. Demidov. "Electrophysical Properties of Superconducting Film–Electrolyte Interface. *Sverchprov. Fiz. Khim. Tekh.* 4:1769–1799, 1991 (in Russian).

[170] A. M. Yafyasov and P. P. Konorov. "Investigation of the H_c-Temperature YBaCuO-Electrolyte Interface. *Sverchprov. Fiz. Khim. Tekh.* 2:133–135, 1989 (in Russian).

[171] P. Boddy and W. Brattain. "The Distribution of Potential at the Germanium–Aqueous Electrolyte Interface." *J. Electrochem. Soc.* 110:570–576, 1963.

[172] O. V. Romanov, M. A. Sokolov, and S. N. Sultanmagomedov. "Potentiodynamic Measuring on the Semiconductor Electrode. Semiconductors Si and Ge. *Elektrokhimiya* 16:935–943, 1980 (in Russian).

[173] H. Gerischer, M. Hoffmann-Perez, and W. Mindt. "Uber den chemischen Zustand und das electrische Moment der Oberflache von

Germanium in kontakt mit wasrigen Electrolytlosungen." *Ber. Bunsenges. Phys. Chem.* 69:130–138, 1965.

[174] V. B. Bogevolnov, A. M. Yafyasov, and P. P. Konorov. "The Charge Dynamics and Self-Organization in a Semiconductor-Electrolyte Interface." *Vestn. Leningr. Univ., Fiz., Khim.* 2(11):29–34, 1993 (in Russian).

[175] V. B. Bogevolnov, A. M. Yafyasov, and P. P. Konorov. "Field-Effect in a Semiconductor-Electrolyte Interface." *Vestn. Leningr. Univ., Fiz., Khim.* 1:18–24, 1993 (in Russian).

[176] R. Memming. "On the Origin of Fast Surface States at the Germanium-Electrolyte Interface." *Surf. Sci.* 2:436–443, 1964.

[177] R. Memming and G. Neumann. "On the Relationship between Surface States and Radicals at the Ge-Electrolyte Interface." *Surf. Sci.* 10:1–20, 1968.

[178] R. Pribil. *Analytical Application of EDTA and Related Compounds.* Pergamon Pres, New York, 1972.

[179] T. Ando, A. B. Fowler, and F. Stern. "Electronic Properties of Two-Dimensional Systems." *Rev. Mod. Phys.* 54:1–225, 1982.

[180] R. Dornhaus and G. Nimtz. In *Narrow-Gap Semiconductors*, 309. Vol. 98 of *Springer Tracts in Modern Physics*. Springer-Verlag, Berlin, 1983.

[181] T. Ando, Y. Arakawa, K. Furuda, S. Komiyama, and H. Nakashima. *Mesoscopic Physics and Electronics*. Springer-Verlag, Berlin, 1998.

[182] "New Technology of Control of Narrow-Gap Semiconductors." Report, EC-ESPRIT Project NTCONGS-28890, 2000.

[183] A. Yacoby, H. L. Stormer, Ned S. Wingreen, L. N. Pfeiffer, K. W. Baldwin, and K. W. West. "Nonuniversal Conductance Quantization in Quantum Wire." *Phys. Rev. Lett.* 77:4612–4615, 1996.

[184] J. R. Meyer, F. J. Bartoli, C. A. Hoffman, and L. R. Ram-Mohan. "Band-Edge Properties of Quasi-One-Dimensional HgTe-CdTe Heterostructures." *Phys. Rev. Lett.* 64:1963–1966, 1990.

[185] Y. Meir, K. Hirose, and N. S. Wingreen. "Kondo Model for the '0.7 Anomaly' in Transport through a Quantum Point Contact." *Phys. Rev. Lett.* 89:1961–1965, 2002.

[186] R. de Picciotto, H. L. Stormer, A. Yacoby, K. W. Baldvin, L. N. Pfeiffer, and K. W. West. "The Influence of Contacts and Inhomogeneities on the Conductivity of Nanowires." *Physica E* 6:514–517, 2000.

[187] K. Tanaka, Y. Nakamura, and H. Sakaki. "Anomalous Conductance in a Novel Quantum Point Contact with Periodic Potential Modulation." *Physica E* 6:558–560, 2000.

[188] T. J. Thomas, J. T. Nicholls, N. J. Appleyard, M. Y. Simmons, M. Pepper, D. R. Mace, W. R. Tride, and D. A. Ritchie. "Interaction Effects in a One-Dimensional Constriction." *Phys. Rev. B* 58:4846–4852, 1998.

[189] A. M. Yafyasov, I. M. Ivankiv, V. B. Bogevolnov, B. S. Pavlov, and T. V. Rudakova. "Mathematical Modelling and Self-Consistent Calculation of the Charge Density of Two-Dimensional Electron's System." *Report Series No. 338*, 1–17. Auckland. New Zealand, June 1997.

[190] A. M. Yafyasov and I. M. Ivankiv. "Self-Consistent Quantum Calculation of Space Charge Region for Accumulation and Inversion Band Bending." *Phys. Status Solidi B* 208:41–49, 1998.

[191] A. M. Yafyasov, I. M. Ivankiv, and V. B. Bogevolnov. "Quantization of the Charge Carriers on InSb at Room Temperature." *Appl. Surf. Sci.* 142:629–632, 1999.

[192] G. Nimtz, J. X. Huang, J. Lange, L. Mester, and H. Spieker. "Electron Wave Interference Effects in CMT and Quantum-Size Devices." *Semicond. Sci. Technol.* 6:C130–C132, 1991.

[193] D. Eger and Y. Goldstein. "Quantization Effects in ZnO Accumulation Layers in Contact with an Electrolyte." *Phys. Rev. B* 19:1089–1097, 1979.

[194] N. N. Ovsiuk and M. P. Sinukov. "Influence of Quantization in Subsurface of Area of the Semiconductor on Spectra of Electroreflection." *Pis'ma Zh. Eksp. Teor. Fiz.* 32:366–370, 1980 (in Russian).

[195] G. P. Panosyan, Z. A. Kasamyan, and G. Sh. Scavenyan. "The Size Quantization of Surface Excitons on the CdTe-Electrolyte Interface." *Fiz. Tekh. Poluprovodn.* 25:1030–1033 1991 (in Russian).

[196] B. N. Zvonkov, B. N. Salashenko, and O. N. Filatov. "The Size Quantization of the InSb Surface." *Fiz. Tverd. Tela* 21:1344–1348, 1979 (in Russian).

[197] A. M. Brodskii, A. Z. Zaidenberg, and A. M. Skundin. "The Metalization and Manifestation of Size Quantization in Bi Film–Electrolyte System." *Pis'ma Zh. Eksp. Teor. Fiz.*, 40:3–5, 1984 (in Russian).

[198] A. M. Brodskii, A. Z. Zaidenberg, and A. M. Skundin. "Photoemission from Size-Quantized Bismuth Films into Electrolyte." In *Proceedings of the International Conference on the Electrodynamics and Quantum Phenomena at Interfaces*, 1–5, Telavi, USSR, 1984.

[199] P. P. Konorov, A. M. Yafyasov, and V. B. Bogevolnov. "Low-Dimensional Effects on the GaSb Semiconductor Surface." *Phys. Low-Dimens. Semicond. Struct.* 2/3:133–138, 1995.

[200] V. M. Asnin, A. A. Rogachev, A. Yu. Silov, and V. I. Stepanov. "Two-Dimensional Electron-Hole Condensate." *Solid State Commun.* 74:404–410, 1990.

[201] V. M. Asnin, A. A. Rogachev, V. I. Stepanov, and A. B. Churilov. "Two-Dimensional Electron-Hole Condensate on the Germanium Surface." *Pis'ma Zh. Eksp. Teor. Fiz.* 43:284–287, 1986 (in Russian).

[202] M. W. Scott. "Electron Mobility in HgCdTe." *J. Appl. Phys.* 43:1055–1062, 1972.

[203] M. Sakahita, B. Lochel, and H. H. Strehblow. "An Examination of the Electrode Reactions of Te, HgTe and $Cd_{0.2}Hg_{0.8}Te$ with Rotating-Split-Ring-Disc Electrodes." *J. Electroanal. Chem.* 140:75–79, 1982.

[204] I. M. Ivankiv, A. M. Yafyasov, and V. B. Bogevolnov. "Method for Determining the Stoichiometric Composition of a Mercury Cadmium Telluride Solid Solution from Capacitance-Voltage Characteristics." *Semiconductors* 35(5):525–528, 2001.

[205] A. Yafyasov, V. Bogevolnov, I. Ivankiv, and O. Shevcenko. "Peculiarities of Electron Effect Effective Mass in Inversion Layers of Kane Semiconductor at Room Temperature." In *Proceedings of the Ninth International Conference on Narrow Gap Semiconductors*, ed. N. Puhlman, H.-U. Muller, and M. von Ortenberg. Humbolt University Press, Berlin, 2000.

[206] D. D. MacDonald. "Applications of Electrochemical Impedance Spectroscopy in Electrochemistry and Corrosion Science. In *Techniques for Characterization of Electrodes and Electrochemical Processes*, ed. R. Varma and J. R. Selman. Willey, New York, 1991.

[207] W. P. Gomes and D Vanmaekelbergh. "Impedance Spectroscopy at Semiconductor Electrodes: Review and Recent Developments." *Electrochim. Acta* 41:967–973, 1996.

[208] V. V. Scherbakov. "Account of Electrical Capacity of a Solution in the Analysis of Impedance of an Electrochemical Cell." *Elektrokhimiya* 34:122–125, 1998 (in Russian).

[209] G. Oskam, D. Vanmaekelbergh, and J. Kelly. "A Reappraisal of the Frequency Dependence of the Impedance of Semiconductor Electrodes." *J. Electroanal. Chem.* 315:65–85, 1991.

[210] D. Mc Donald, E. Sikora, and G. Engelhardt. "Characterizing Electrochemical Systems in the Frequency Domain." *Electrochim. Acta* 43: 87–107, 1998.

[211] F. Mansfeld, C. Lee, and G. Zhang. "Comparison of Electrochemical Impedance and Noise Data in the Frequency Domain." *Electrochim. Acta* 43:435–438, 1998.

[212] V. V. Pototskaja and N. E. Evtushenko. "Frequency Dependence of Gerischer's impedance on an Electrode with a Modelling Surface." *Elektrokhimiya* 34:513–519, 1998 (in Russian).

[213] M. D. Krotova and Yu. V. Pleskov. "On the Measurements of the Surface Conductance of Germanium in Aqueous Solutions." *Phys. Status Solidi* 3:2119–2126, 1963.

[214] H. Von Gobrecht und O. Meinhardt. "Uber die Impedanz von Halbleiter-Electrolyt-Grenzfdlachen. Mitt III." *Ber. Bunsenges. Phys. Chem.* 67:151–160, 1963.

[215] H. Von Gobrecht, O. Meinhardt, und B. Reinicke. "Uber die Einfluss-Ion Oberflachen Zustanden auf das electrische Verhalten von

Germanium und Silizium Electroden." *Ber. Bunsenges. Phys. Chem.* 67:493–500, 1963.

[216] O. V. Romanov, M. A. Sokolov, and M. S. Griliches. "To a Question on Measurements of Conductivity Germany in Electrolytes." *Elektrokhimiya* 10:584–586, 1974 (in Russian).

[217] S. Y. Wang, F. Haran, J. Simpson, H. Stewart, J. Wallace, K. Prior, and B. Cavenett. "Electrochemical Capacitance-Voltage Profiling of n-Type Molecular Beam Epitaxy ZnSe Layers." *Appl. Phys. Lett.* 60:344–346, 1992.

[218] V. I. Shashkin, I. P. Karetnikova, A. V. Murel, I. M. Nefedov, and I. A. Shereshevski. *Fiz. Tekh. Poluprovodn.* 31:926, 1997 (in Russian).

[219] P. Blood. "Capacitance-Voltage Profiling and the Characterisation of III-V Semiconductors Using Electrolyte Barriers." *Semicond. Sci. Technol.* 1:7–27, 1986.

[220] M. A. Abaturov, V. V. Bakin, and M. D. Krotova. "Definition of Concentration and Profile of Concentration of the Doping Impurity in Semiconductor Electrodes." *Elektrokhimiya* 31:1214–1220, 1995 (in Russian).

[221] M. M. Faktor and J. L. Stevenson. "The Detection of Structural Defects in GaAs by Electrochemical Etching." *J. Electrochem. Soc.* 125:621–628, 1978.

[222] T. Ambrigde, J. Stevenson, and M. Redstall. "Applications of Electrochemical Methods for Semiconductor Characterization." *J. Electrochem. Soc.* 127:222–227, 1980.

[223] M. Buda, E. Smalbrugge, E. Geluk, F. Karouta, G. Acket, T. vande Roer, and L. Kaufman. "Controlled Anodic Oxidation for High Precision Etch Depth in AlGaAs III-V Semiconductor Structures." *J. Electrochem. Soc.* 145:1076–1080, 1998.

[224] V. B. Bogevolnov and A. M. Yafyasov. "Charge Nature of Stabilization of Electric Potential inside a Cell Culture." *Vestn. St. Petersburg Univ., Fiz., Khim.* 1(4):118–120, 2004 (in Russian).

[225] Alan Marty and Erwin Neher. "Whole-Cell Recording." In *Single-Channel Recording*, ed. Bert Sakmann and Erwin Neher. Plenum Press, New York, 1986.

[226] Yasouo Kagava. *Biomembranes*, translated from Japanese. Vissh. Shkola., Moscow, 1987 (in Russian).

[227] P. K. Anochin. "Theory of Functional Systems as a Prerequisite for Creation of Biocybernetics." *Biologic Aspects of Cybernetics*, 74–91. Nauka, Moscow 1962.

[228] P. P. Konorov and A. M. Yafyasov. *The Physics of Semiconductor Electrode Surfaces.* S. Petersburg University Press, Saint Petersburg, 2003 (in Russian).

[229] M. J. Weaver and X. Gao. "In Situ Electrochemical Surface Science." *Annu. Rev. Phys. Chem.* 44:459–494, 1993.

[230] J.-H. Chazalviel. "Experimental Techniques for the Study of the Semiconductor-Electrolyte Interface." *Electrochim. Acta* 33:461–476, 1988.

[231] D. Kolb. Chap. 4 in *Spectroelectrochemistry: Theory and Practice*, ed. R. J. Gale, 87–188. Plenum Press, New York, 1988.

[232] B. Erne, M. Stchakovsky, F. Ozanam, and J.-H. Chazalviel. "Surface Composition of n-GaAs Cathodes during Hydrogen Evolution Characterized by in situ Ultraviolet-Visible Ellipsometry and in situ Infrared Spectroscopy." *J. Electrochem. Soc.* 145:447–456, 1998.

[233] S. J. Higgins, P. A. Christensen, and A. Hamnett. "In situ Ellipsometry and FTIR Spectroscopy Applied to Electroactive Polymer-Modified Electrodes. Chap. 2 in *Electroactive Polymer Electrochemistry*, ed. M. E. G. Lyons. Part 2, 132–137. Plenum Press, New York, 1996.

[234] J. Pemberton and S. Garvey. "Spectroscopic Methods for the Characterization of Electrochemical Interfaces and Surfaces." Chap. 2 in *Modern Techniques in Electroanalysis*, ed. P. Vanysek, 59–106. Wiley, New York, 1996.

[235] W.-K. Paik. "Ellipsometry in Electrochemistry." *Mod. Aspects Electrochem.* 25(4):191–252, 1993.

[236] J. Rupich, V. Timoshenko, and V. Dittrich. "In situ Monitoring of Electrochemical Processes at the p-Si (100)/Aqueous NH_4F Electrolyte Interface by Photoluminescence." *J. Electrochem. Soc.* 144:493–500, 1997.

[237] S. Gottesfeld, Y.-T. Kim, and A. Redondo. "Recent Applications of Ellipsometry in Electrochemical Systems. Chap. 9 in *Physical Electrochemistry: Principles, Methods and Applications*, ed. I. Rubinstein, 393–467. Marcel Dekker, New York, 1995.

[238] J. McBreen. "In situ Synchrotron Techniques in Electrochemistry." Chap. 8 in *Physical Electrochemistry: Principles, Methods and Applications*, ed. I. Rubinstein, 339–391. Marcel Dekker, New York, 1995.

[239] F. Kong, J. Kim, X. Song, M. Inaba, K. Kinoshita, and F. McLarnon. "Exploratory Studies of the Carbon/Nonaqueous Electrolyte Interface by Electrochemical and in situ Ellipsometry Measurements." *Electrochem. Solid State Lett.* 1:39–41, 1998.

[240] A. Chaparro, P. Salvador, and A. Mir. "The Scanning Microscopy for Semiconductor Characterization: Photocurrent, Photovoltage and Electrolyte Electroreflectance Imaging in the n-MoSe$_2$/I Interface." *J. Electroanal. Chem.* 424:153–158, 1997.

[241] J. Ruppich, V. Timoshenko, and V. Dittrich. "In situ Monitoring of Electrochemical Processes at the (100) p-Si/Aqueous NH_4F Electrolyte

Interface by Photoluminescence." *J. Electrochem. Soc.* 144:493–497, 1997.

[242]   J. N. Chazalviel, C. da Fonseca, and F. Ozanam. In situ Infrared Study of the Oscillating Anodic Dissolution of Silicon in Fluoride Electrolytes." *J. Electrochem. Soc.* 145:954–974, 1998.

[243]   Z. Deng, J. Spear, and J. Rudnicki. "Infrared Photo Thermal Deflection Spectroscopy—A New Probe for the Investigation of Electrochemical Interfaces." *J. Electrochem. Soc.* 143:1514–1521, 1996.

[244]   A. Tadjeddine, A. Peremans, and P. Gayot-Sionnest. "Vibrational Spectroscopy of the Electrochemical Interface by Visible-Infrared Sum Frequency Generation." *Surf. Sci.* 335:210–220, 1995.

[245]   *Infrared Spectroscopy in Electrochemistry*. Special issue, *Electrochim. Acta* 41(5), 1996.

[246]   C. Korzeniewski. "Infrared Spectroscopy in Electrochemistry: New Methods and Connections to UHV Surface Science." *Crit. Rev. Anal. Chem.* 27:81–102, 1997.

[247]   T. Iwasita and F. C. Nart. "In-situ Infrared Fourier Transform Spectroscopy: A Tool to Characterize the Metal-Electrolyte Interface at a Molecular Level." *Adv. Electrochem. Sci. Eng.* 4(3):123–216, 1955.

[248]   J.-N. Chazalviel, F. Ozanam, and A. Djebry. "Hydratation Si Surface in Organic Electrolytes Probed Through in situ IR-Spectroscopy." *Electrochim Acta* 41:687–692, 1996.

[249]   H. D. Abruna. "X-ray Absorption Spectroscopy in the Study of Electrochemical Systems." Chap. 1 in *Electrochemical Interfaces: Modern Techniques for in-situ Interface Characterization*, ed. H. D. Abruna, 1–54. VCH, New York, 1991.

[250]   A. Mansour, C. Melendres, and J. Wong. "In situ X-Ray Adsorption Spectroscopic Study of Electrodeposited Nickel Oxide Films during Redox Reactions." *J. Electrochem. Soc.* 145:1121–1126, 1998.

[251]   G. G. Long and J. Kruger. "Surface X-Ray Absorption Spectroscopy, EXAFS and NEXAFS, for the in situ and ex situ Study of Electrodes." Chap. 4 in *Techniques for Characterization of Electrodes and Electrochemical Processes*, ed. R. Varma and J. R. Selman, 167–209. Wiley, New York, 1991.

[252]   H. D. Abruna. "Probing Electrochemical Interfaces with X-Rays." *Adv. Chem. Phys.* 77:255–335, 1990.

[253]   C. Melendres and A. Mansour. "X-Ray Absorption Spectroelectrochemical Cell for "in situ" Studies of Thin Films." *Electrochim. Acta* 43:631–634, 1998.

[254]   E. Meulenkamp. "An Electrochemical Cell for Simultaneous in situ X-Ray Diffraction and Optical Measurements." *J. Electrochem. Soc.* 145:2759–2763, 1998.

[255] D. Lynch. "Application of Synchrotron Radiation to the Study of the Solid-Electrolyte Interface." *J. Electroanal. Chem.* 150:293–333, 1983.

[256] W. Paatsch. "Investigation of Passive Electrodes Using Modulation Spectroscopy and Photo-Potential Measurements." *J. Phys.* 38: 151–155, 1977.

[257] R. Mc Intyre and J. Sass. "Inverse Photoemission Spectroscopy at the Metal-Electrolyte Interface." *Phys. Rev. Lett.* 56:651–659, 1986.

[258] C. Melenders and A. Mansour. "X-ray Adsorption Spectroelectrochemical Cell for in-situ Studies of Thin Films." *Electrochim. Acta* 43: 631–634, 1998.

[259] K. Hirasawa, T. Sato, and S. Yamaguchi. "Force Microscope Study on Graphite Electrodes." *J. Electrochem. Soc.* 144:L81–L84, 1997.

[260] R. Sonnenfeld and P. Hansma. "Atomic-Resolution Microscopy in Water." *Science* 232:211–213, 1986.

[261] S. R. Higgins and R. J. Harners. "Spatially Resolved Electrochemistry of the Lead Sulfide (Galena) (001) Surface by Electrochemical Scanning Tunneling Microscopy." *Surf. Sci.* 324:263–281, 1995.

[262] A. J. Bard, F.-R. F. Fan, and M. V. Mirkin. "Scanning Electrochemical Microscopy." *Electroanal. Chem.* 18(3):244–373, 1994.

[263] M. Vela, G. Andreasen, S. Aziz, R. Salvarezza, and A. Arvia. "Sequential in situ STM Imaging of Electrodissolving Copper in Different Aqueous Acid Solutions." *Electrochim. Acta* 43:3–12, 1998.

[264] R. Sonnenfeld, J Schneir, and P. K. Hansma. "Scanning Tunneling Microscopy: A Natural for Electrochemistry." *Mod. Aspects of Electrochem.* 21(1):1–28, 1990.

[265] A. A. Gewirth and B. K. Niece. "Electrochemical Applications of in situ Scanning Probe Microscopy." *Chem. Rev. (Washington, D.C.)* 97:1129–1162, 1997.

[266] M. Ward and H. White. "Scanning Tunneling and Atomic Force Microscopy of Electrochemical Interfaces." Chap. 3 in *Modern Techniques in Electroanalysis*, ed. P. Vanysek, 107–150. Wiley, New York, 1996.

[267] X. Jiang, M. Sato, and N. Sato. "Piezoelectric Detection of Oxide Formation and Reduction on Pt Electrodes." *J. Electrochem. Soc.* 138:137–139, 1991.

[268] J. Lorimer, B. Pollet, S. Phull, T. Mason, and D. Walton. "The Effect upon Limiting Currents and Potentials of Coupling a Rotating Disc and Cylindrical Electrode with Ultrasound." *Electrochim. Acta* 43:449–455, 1998.

[269] S. Ching and J. McDevitt. "Liquid Phase Electrochemistry at Ultralow Temperatures." *J. Electrochem. Soc.* 138:2308–2315, 1991.

[270] D. H. Evans and S. A. Lerke. "Electrochemical Studies at Reduced Temperature." Chap. 16 in *Laboratory Techniques in Electroanalytical*

Chemistry, ed. P. T. Kissinger and W. R. Heineman, 487–510. Dekker, New York, 1996.

[271] T. Iwasita, S. Rottegerman, and W. Schmickler. "The Double-Layer Capacity in Liquid and Solid Aqueous Electrolyte." *J. Electroanal. Chem.* 196:203–209, 1985.

GPSR Authorized Representative: Easy Access System Europe - Mustamäe tee
50, 10621 Tallinn, Estonia, gpsr.requests@easproject.com